THE OLIVE BOOK

THE OLIVE BOOK

Gareth Renowden

CANTERBURY UNIVERSITY PRESS

First published in 1999 by
CANTERBURY UNIVERSITY PRESS
University of Canterbury
Private Bag 4800
Christchurch
NEW ZEALAND
mail@cup.canterbury.ac.nz
www.cup.canterbury.ac.nz

Reprinted 2000

ISBN 0-908812-80-9

Cover photograph by G.R. 'Dick' Roberts, Nelson.
All photos by the author unless otherwise credited.

Designed and typeset by Mark Winstanley / Go Ahead Graphics
Printed in New Zealand by The Caxton Press, Christchurch

CONTENTS

1	The Olive Tree	9
2	An Olive Primer	16
3	Olives Around the World	30
4	From Tree to Table	43
5	Tastings and Using Olive Oil	49
6	Getting Started	55
7	Financial Planning	65
8	Selecting Your Trees	71
9	Planting the Grove	87
10	Grove Management	100
11	Pest, Diseases and Other Problems	111
12	Oil Production	118
13	Olives for Eating	128
14	Olive Cuisine	133
15	Books, Contacts and Other Sources	140
Index		143

For Camille, Tim and Emma.

Acknowledgements

In researching this book, I was continually impressed by the helpfulness and generosity of the "olive people" I met. From deepest New South Wales to sunny Marlborough, I was always met with a smile and a willingness to shed a little light into the darkness of my ignorance. I'd like to thank the following people for all their help: in New Zealand, Hamish McFarlane and the team at Marlborough Olives, Tony Harvey at Ponder Estate, Brian Bicknell at Seresin Estate, Geoff Elliot of Elliot's Wholesale Nursery in Amberley, Jill Moore at Waipara Springs, and Mike and Jane East at Muddy Water vineyard. Special thanks to Roland Clark ("Norwester") for loaning books, pictures, and an always informed view. In Australia, I must thank Ann and Kent Hallett for their generosity in both opening doors for me in South Australia and providing me with a bed while I explored the region. Margaret Kirkby and her family also proved that Aussie hospitality is a wonderful thing. I'd like to thank Jenny Birch, Ian Wolfgang, Michael Burr, Peter Maroudas, Colin Longmire, Ben Walker, Louise Jacka, Jim Smyth, Andrew Burgess and Julian Archer at Australian Olives, and Penny Hart.

The team at Canterbury University Press also deserve a mention, not least Mike Bradstock for being brave enough to commission this book. Sandra Parkkali produced some great illustrations, Mark Winstanley produced an elegant design, and Anna Rogers edited with finesse.

Finally, thanks to my family for putting up with my tetchiness when deadlines loomed, and for being prepared to share my vision.

PREFACE

When I wanted to find a book that would guide me through planting some olive trees and creating oil, I couldn't find one. There were technical manuals, there were 'foodie' titles to help you appreciate fine oils, and there were stacks of recipe books based on olives and oil, but there was nothing that could take a novice through the process of buying, planting, growing, harvesting and pressing or pickling the fruit. So I decided that if I was going to have to do all the research to get my own trees into the ground, then I might as well share it with the world, and I was fortunate enough to find a publisher who agreed.

Before my wife and I left a West London suburb a few years ago for the South Island of New Zealand – the land of my wife's fathers – we did a number of things that we'd been meaning to do for a while. One of these was a holiday in Tuscany. Tasting oils in Chianti – especially at the old abbey at Coltibuono (which also has a fabulous restaurant) – imprinted itself on my palate, as well as on the palates of several friends who benefited from a bottle on our return.

It took a while to get acclimatised after arriving in New Zealand, but we slowly began to realise what was possible with the proceeds from selling a modest London home. A chance encounter with a house for sale with truffle plantation attached, led to research about the feasibility of growing truffles and, by a serendipitous trail, to a farmhouse and 10 hectares of land it would have been folly to leave. With memories of Tuscany fresh in my mind, and data suggesting that our corner of North Canterbury had the same sort of climate as the South of France and Northern Italy, it wasn't long before I began to add an olive grove to my plans for a truffière.

In my naive way, I thought I could set about recreating some of the fine oils I'd enjoyed. And so I walked into the Canterbury Library and demanded everything they had on olives. It was a short list.

This book, then, is for anyone fortunate enough to live in an area that will grow good olives, and who wishes to try their hand at planting some trees and dealing with the fruit. It is both a record of my research and a practical guide to getting started. It is not the result of 40 years of apprenticeship to a grizzled old Tuscan olive grower, nor is it the definitive guide to all aspects of the business. But it will set you on your way. And if you find the whole business as fascinating as I do, you will become, and remain, hooked on exploring the mysteries of the olive tree and its fruit.

THE OLIVE TREE

Olea europea*, the European olive. A tough tree, used to hot summers and long periods of low rainfall, yet capable of growing to 10 metres in height and living for a very long time. It is said that the olives in the Garden of Gethsemane are the same ones that were growing in the time of Christ. Their silver and green leaves flicker attractively in the wind and provide perfect shade: just the right mixture of sun and air. The ripe fruit is inedibly bitter, but rich in oil – the only vegetable oil that can be extracted by a simple process and used without further treatment.

The olive and its oil underpin the development of human civilisation. Persians, Egyptians, Cretans, Greeks and Romans loved the stuff, made it the basis of their economies. It was burned in lamps to free humanity from the tyranny of the dark. When perfumed, it adorned the bodies of the wealthy and the athletic, and in food it provided the energy that kept life moving. It became interwoven with the daily and spiritual lives of peoples throughout the Mediterranean, becoming a symbol of peace and redemption, of fruitfulness and virtue. And then, when the centre of civilisation moved northwards, to the cold lands where animal fats replaced olive oil and it was impossible to grow the tree, the olive became all symbol and no substance. When the northern Europeans expanded around the world, taking with them their foods and habits, industrialising and modernising and sanitising as they went, the olive and its oil were cherished only in their heartland.

In the modern world, the olive is resuming its rightful place as the basis of a diet for a healthy life. Fine olive oil is a gourmet delight, and widely touted as beneficial to health. Pickled olives in a myriad of forms decorate drinks, canapés, pizzas and fashionable restaurant dishes. In the Mediterranean and the cuisines of the South of France, Spain, Italy, Greece and the Middle East and North Africa, the olive never went away. In the Southern Hemisphere, the olive accompanied settlers wherever they went, arriving in the Americas, South Africa, Australia and New Zealand soon after the first Europeans. And it is in these countries that olive growing is now taking hold in a big way. In Australia, growers confidently expect to be able to fulfil local demand for quality oil early in the new millennium, and soon thereafter export to the world. In Argentina, massive new

The gnarled old olive trees in the Garden of Gethsemane may well have been around in the time of Christ. *Roland Clark*

plantations are being established, and New Zealand growers are beginning to produce oils that compare with the world's best.

Why this upsurge of interest? The reasons vary from country to country, from simple economics to a sentimental desire to connect with the land, but at the root of all this expansion is the recognition that the 'Mediterranean diet' is both delicious and good for you. And a key part of that diet is the use of olive oil. Throughout the Western world, consumers are becoming more concerned about the food they eat and its impact on their health. In the United States, the desire to postpone the effects of age and to maintain a healthy body is more than just fashion. As demand for the good oil rises, so does the appreciation of the best olive oils as a definite gourmet experience. Borrowing the trappings of wine tasting, people are now asked to compare oils, sniffing and slurping until the distinctive flavours ignite on the palate.

With fashion, comes sentiment. Millions read books about seasons in Tuscany, houses in Provence, or return from Mediterranean holidays with the fuzzy, feel-good idea that 'a few olives would be nice'. Somehow, through the harsh economic and social realities of the turn of the century, the olive retains its resonance and romantic appeal, especially for those who can afford to dabble in 'lifestyle' farming – or who just have a large garden.

Other farmers look to olives for more agronomically sound reasons: diversification, achieving sustainable land use, or as a response to threats of lower rainfall caused by climate change. Low prices for many agricultural products lead to a willingness to explore alternatives. In many parts of Australia and New Zealand, the returns from olive growing could be better than other, more traditional land uses such as pastoralism, and much more environmentally friendly. If your pasture is becoming marginal because of prolonged or more frequent droughts, then a covering of hardy trees is certainly an attractive alternative; you may even be able to graze sheep in the shade of the olives, as the Greeks do.

When we look at a piece of land in Australia or New Zealand, with its own unique combination of soil and climate – what French winemakers would call *terroir* – we have too little history to have much idea about what the best agricultural use might be. In Italy or France, when you look at a landscape, you see a patchwork of agricultural uses developed over thousands of years of trial and error. The pioneers of grape growing in the Wairau Valley in Marlborough, New Zealand were backing a strong hunch

Tuscan trees growing under a southern sun. Marlborough's Seresin Estate is hoping to produce oil to match the world's best.

– but that was only 20 years ago. Look at the vineyards in Burgundy, at the famous Côte d'Or, and you'll see how centuries of experience have refined the understanding of the best sites for ripening the pinot noir grape – and how that is reflected in the price of the wine from each of the many small vineyards.

The new Southern Hemisphere growers are going to change the face of the olive world. The wine business set the precedent, producing respectable wines that could look the European aristocracy in eye, and then unleashing sauvignon blancs that took the world by storm. The traditional olive growing areas of the Mediterranean and Middle East will have to face an assault on two fronts. The bulk growers of Australia and Argentina will produce very high quality olive oil at reasonable prices and, despite having to compete with heavily subsidised European growers, should soon have an impact on the world market. Smaller growers, aiming for the highest quality, experimenting with varieties and styles, will create oils that could worry the stars of the Tuscan hills. As the best of the traditional olive varieties are planted in the various climates of Australia and New Zealand, there is a good chance that a whole new style of oil could emerge.

Nothing will ever be simple, of course, when you are dealing with an agricultural product at the mercy of the climate and a fickle world market. The wrong sorts of olives will be planted, trees will die, growers and processors will get into financial difficulties, and there may even be a bust to follow the boom.

But there's still a romance about olives, even when you're talking about planting thousands of trees to be mechanically harvested with not a gnarled peasant in sight. Every grower I've met has been an enthusiast, delighted to meet someone of like mind, willing to share experiences and expertise, discuss problems, seek advice and, above all, show off their trees. This enthusiasm, and the sheer scale of the investments being made, especially in Australia, mean that this olive boom is unlikely to fizzle out.

REALITY

So what are the realities of establishing an olive grove? (I prefer the term grove, but orchard is just as good, and probably more widely used.) What workload and financial commitment is involved in planting olives to generate some sort of income?

Setting aside all the detailed planning you need to do (and there's a lot – more than you may expect), the preparation and planting of even a modest olive grove represents a substantial workload. If you intend to do it yourself, be prepared for an aching back. There's irrigation to arrange, rows to lay out and mark, weeds and grass to spray off, the 'ripping' of the rows with a big metal hook to make it easy for the roots to grow, compost or fertiliser to distribute, stakes to bang into the ground, and then loads of holes to dig for the little trees. Next you have to put the trees into the holes, firm the soil around them, protect them from wind and pests (ranging from rabbits to kangaroos and cockatoos, depending on where you are) and turn on the irrigation. Then you pour a large glass of red wine, admire the view and take the next day off to recover.

With the trees in the ground, you can relax a bit. For the first couple of years you need to monitor the health of the trees, looking for nasty bugs and mineral deficiencies, and making sure that they don't die in the hardest frost for 40 years, but there's not much pruning to be done, nor any fruit to harvest. By years three and four, you'll be starting to prune the trees to establish the right shape. Count on several minutes per tree, then multiply

A five-year-old Kalamata tree thriving in the northwest New South Wales climate.

by the number of trees, and realise that you face a major task. The first fruit should be appearing, but probably only in quantities suitable for pickling and showing off to your friends.

By year five, you should be getting some fruit – enough to be worth picking. Olives ripen in autumn and winter, so the cooler your climate, the later they'll ripen. You'll find yourself picking in the depths of winter, cold and wet, with a sore back and painful fingers. Only then can you get ready for pickling or oil-pressing.

As the trees get bigger, your grove will become much more impressive – and so will the pruning load and picking time. Of course, all this effort will be rewarded by fine oil, spectacular pickled olives and a nice little income. Or it may not. At this stage in the evolution of the business, olive growing cannot be recommended as an easy way to fill up a lifestyle block, or as a guaranteed way of making money. Other tree crops would certainly offer a less labour-intensive alternative. But then picking up chestnuts with a giant vacuum cleaner doesn't have the same appeal, does it?

Once picked, your fruit will need processing – and quickly – if it is to yield a high quality product. It may be all right for European peasants to leave their olives lying around for weeks in musty old sacks waiting to be pressed, but if you do that your oil is going to be fit only for burning in Roman-style lamps. So, you've spent all day in the grove picking the precious fruit; now you have to get it to the press (or press it yourself). The closer the press is to your grove, the easier life will be. If it's a long way away, you need to think about refrigerated transport. If you have your own press, life will be simpler, but not easier, because pressing olives is as much art as science.

If your grove is of any size, you'll end up with lots of large tubs full of pickled olives, or barrels of olive oil. And then you have to sell it. This is no trivial matter. Friends may tell you that the oil is wonderful, pester you for loads of free bottles, but that doesn't get labels designed, bottles bought and filled, retailers persuaded to order and stock (and pay for) the product. Word of mouth may be the best form of recommendation, but it's a pretty slow and fickle method of mass marketing.

You will also have to comply with all the local food hygiene requirements, keeping things scrupulously clean. Good insurance might be useful. One of the possible problems with pickled olives is the production of the botulinus toxin, a rather nasty poison.

AN OLIVE PRIMER

Whether you want to grow olives on a commercial basis, or need to know more about the trees in your garden, it helps to understand the botanical basics of *Olea europea* and something of its history and health benefits.

BOTANY

The olive is a tree that wants to be a shrub, or a shrub that can be persuaded to be a tree, and occurs in the wild around the Mediterranean basin. The wild olive, or *Oleaster*, is one of up to 29 members (genera) of the family Oleaceae (depending on which expert you choose to believe), that also includes things like *Syringa* (lilacs), *Jasminus* (jasmine), *Fraxinus* (ash) and *Forsythia*. The species *Olea*, which contains *Olea europea*, the cultivated olive, can be divided up in many different ways (those experts again), mainly because it is very difficult to distinguish between 'wild' and 'escaped' olives. In fact, some experts think that many *Olea* species may be what are called 'ecotypes' or 'eco-subspecies' – the same basic tree responding to different climates and soils by adopting a different shape or size. To complicate things even further, some authorities believe there may be as many as 2000 cultivated varieties, or cultivars, in groves around the world, and seedlings from these have inevitably escaped into the wild.

Left to its own devices, *Olea europea* will grow into a 10-metre tree with a round, dense crown. The leaves are spindle-shaped, dark green on the top and silvery underneath. It flowers in late spring, producing clusters of small white inflorescences, mainly on last year's growth. Flowering needs light to get started, so flowers and fruit are carried only on the outside of the tree, unless it is pruned to let light into the middle of the canopy. The fruit matures in late autumn and winter, starting out green and then gradually moving through a reddish purple to black.

Seedlings have a juvenile phase with distinctly different leaves – small and round – but trees produced by rooting cuttings, which is standard nursery practice, don't show this phase unless they are cut right back to ground level.

Olives adapt well to climatic extremes such as drought and heat, and will also tolerate moderate cold. They prefer aerated soils, but will cope in a wide range of soil conditions. A tree's size and fruiting potential depends on its environment. Given good soil, regular feeding and enough water, it will grow quickly and yield heavy crops. But the olive tends to have 'on' and 'off' years. In a year when the tree is carrying a lot of fruit, less energy is put into growing new wood – and since that wood carries the next year's flowers, the following season will see a smaller number of flowers and less fruit. The tree can then focus on putting on lots of new growth, leading to more flowers in the following year, and so the cycle reinforces itself. The pattern is usually much less marked in younger trees, and in groves in

A fine old olive tree provides elegant shade in South Australia's Clare Valley.

cooler areas, but in warmer areas one adverse season can synchronise all the trees in a whole region

The root system of an olive tree depends on whether it's a seedling or a cutting. Seedlings start out with a central leading root that grows straight down, and develop lateral roots only after four or five years – unless transplanted, when they will immediately start to put out laterals. Cutting-grown trees only ever produce lateral roots. In deep and well-aerated soils, the root system can be narrow and go down to 6 or 7 metres, but in shallower soils the roots will spread out, and they can go a very long way. In an irrigated olive grove, most of the roots will be concentrated in the top 70 to 80 centimetres of the soil, with scattered roots down to 1.5 metres or so.

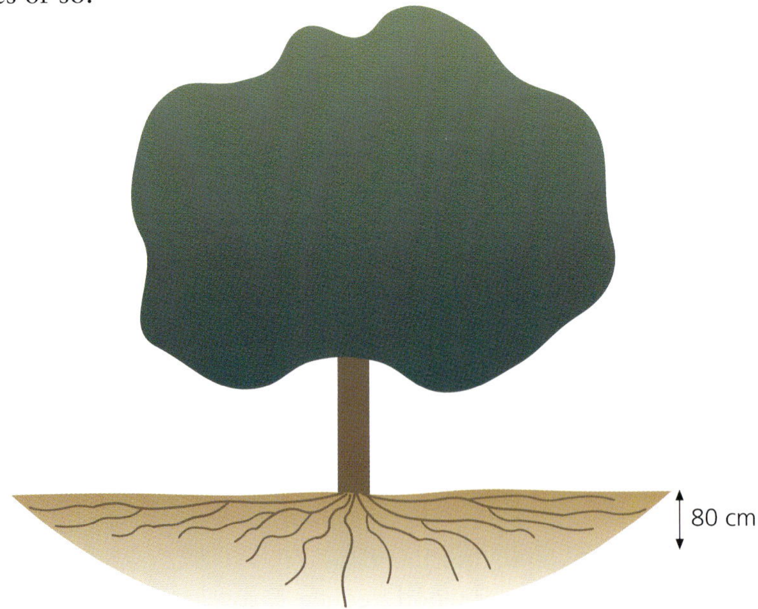

The root system is usually shallow and spread well out beyond the drip line, with the majority of the system concentrated in the top 80cm of the soil.

The root structure is mirrored in the crown of the tree, so if there is poor soil on one side, the top of the tree may be correspondingly smaller or less vigorous on that side. The trunk is basically a bundle of fibres, each connecting one part of the roots to one part of the crown. In older trees, the trunk can develop a distinct 'fluting', as there's less growth where the bundles of fibres touch.

Extensive suckering around the base of the tree has been trimmed away, but there is still plenty of obvious growth going on.

The tree responds very well to pruning, generating new growth very rapidly. The drawback is that an olive tends to shoot from the bottom of any branch, because it is a basally dominant, as opposed to an apically dominant species which will automatically grow nice and straight. Suckers develop rapidly at the base of the trunk, and also from swellings on the trunk known as spheroblasts. If the suckers are not cut off, the tree will rapidly develop a bushy shape. The bark of irrigated trees is much smoother and thinner than the thick, corky bark of dryland trees.

As the tree ages, the inner wood frequently rots away to leave a hollow trunk. This doesn't affect productivity, as long as the canopy is kept youthful by regular pruning, but can weaken the tree and leave it vulnerable to wind damage and breakage. In Europe, few 200 to 300-year-old trees have their original trunks, which is one reason why it can be difficult to work out their precise age. If the original trunk does break off (or is cut off after frost damage, for example), the root system will rapidly push out new growth, and the tree may develop multiple trunks.

Although the adult olive leaf shape does vary from cultivar to cultivar, the differences are slight. Some cultivars, such as Kalamata, are fairly distinctive but, to the untutored eye, the rest look more or less identical.

Kalamata leaves have the most distinctive shape of any cultivar.

Leaves generally live for about two years, dropping when the tree is putting on new growth, or when they get shaded. The leaf structure is very well adapted to conserving moisture, especially on the underside. The size of each leaf varies, depending on the tree's growth cycle; the largest leaves grow when the tree is growing most vigorously – usually in spring.

Flowers normally appear on last year's growth. The flower buds begin to form in the previous summer, and develop over winter inside the new wood. Light stimulates bud formation, and a certain amount of winter cooling is necessary to finish the process. They appear in late spring, the exact timing depending on the local climate – from October to early December in the Southern Hemisphere.

The flowers are mostly wind-pollinated, and though some insects, including bees, may visit the trees, they are not thought to play a very big role. Pollen can travel for very long distances – as much as 7 kilometres has been measured in Israel – but in a new grove you will need to put the trees a little closer together. Cultivars are often self-sterile, needing pollen from a different cultivar to get a good fruit-set. New groves are therefore normally planned with around 10 percent of the trees as pollinators, and the standard advice is to plant as a wide a range of cultivars as possible. You'll need to take the wind direction during the flowering season into account when planning the layout of the grove, especially if your site is prone to prolonged periods of wind from one direction. Hot, dry winds during flowering can affect the fruit-set.

The annual growth cycle of the tree depends on the local climate. In spring, as soon as the temperature climbs above 12°C, growth gets under way, but it can also be switched off again by prolonged periods in summer with temperatures over 30°C. In cooler climates, the tree shows rapid growth in spring, with a summer peak in activity, followed by a decline into winter dormancy. In hotter areas, the olive will 'switch off' in the heat of mid-summer, and resume growth again in the autumn, giving two peaks

Flower buds on a young Manzanilla tree in Canterbury.

to the growth curve. In either case, the availability of water will have a marked influence on the tree's activity.

A certain amount of winter chilling is required to encourage flowering, though the amount required varies from cultivar to cultivar. Temperatures of 5° to 7°C as a part of the normal daily range are said to be the most efficient at inducing flowering, but a steady 12°C has also been shown to be effective. In hot areas, a warm winter may be enough to inhibit flowering in the following year.

Fruit-set is affected by climate and weather, as well as the efficiency of pollination. Within 18 days of full bloom, around 60 percent of the flowers will have been fertilised, and fruit growth will have begun. By 25 days after full bloom, the number of little fruits will have stabilised, and the fruit will begin to turn dark green. Some cultivars are prone to producing 'shotberries', little fruits that never grow to maturity, perhaps caused by early fruitlets inhibiting the growth of relative latecomers.

The fruit goes through five growth stages, the length of each depending on cultivar and climate. The most rapid growth occurs during the second

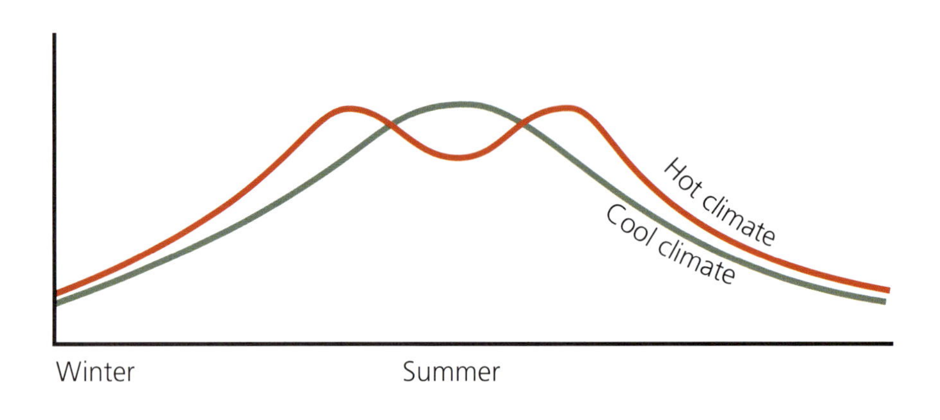

In hot climates, olives will shut down in the midsummer heat, concentrating their growth in spring and autumn spurts.

Fruit can be green ripe as well as black ripe, and ripeness may vary considerably over the canopy of the tree.

phase, from approximately 40 to 60 days old, when the stone is being formed. At the end of this stage, the fruit cells stop dividing and growth occurs by each cell getting bigger. If there is too much stress on the tree during this stage, the overall size of the fruit may be reduced. On the other hand, a period of water stress late in this second stage can reduce the size of the stone, and therefore increase the flesh-to-pit ratio, a useful trick if you're growing olives for the table.

The fruit has a second phase of rapid growth from about 80 to 100 days, when the flesh develops and oil accumulation begins. Giving the tree water during this stage should help to increase the size of the fruit, but you have to be careful not to water too much too early in the year, as this will encourage shoot rather than fruit growth.

The precise time when the fruit ripens depends on the climate and weather in any given season, but is also affected by the fruit load. Fruit on a heavily laden tree may be slower to ripen, and smaller than usual, so thinning earlier in the year may be necessary to maintain fruit size if you're after table olives. The olives are said to be 'green ripe' a few days before the skin of the fruit begins to change colour. The fruit loses some of its firmness to the touch, and in cultivars with a free stone, the stone can be pushed out of the flesh quite easily. As the skin changes colour from green through red/purple to black, oil continues to accumulate in the flesh, but the rate slows down after the whole surface of the fruit is black. The colour change continues in the flesh of the fruit, eventually becoming black all the way through to the stone. The fruit then begins to lose water, and the oleuropein in the flesh, which makes it so bitter, begins to break down. Once all the water has gone, the fruit may be edible – and will certainly be attractive to birds.

HISTORY

Olives were already being used in parts of the Middle East at least 12,000 years ago, but it is thought that the active cultivation of the tree didn't begin until about 6000 years ago. Olive oil was being produced on a large scale in Syria around 3000 BC, and the tree is thought to have spread from there to Egypt. Tablets found in Syria show that, around 1500 BC, olive oil was five times more expensive than wine and two and a half times the price of sesame or linseed oil. By 1200 BC, olive cultivation was widespread in Egypt and the oil was being used in lamps in temples and the homes of the wealthy, and as an ointment to protect the skin from sunburn or cracking. It is still used to soften hard skin today. The oil was already being widely traded around the Middle East, and by ship as far as Cyprus.

The olive is mentioned many times in the Bible, perhaps most famously when the dove returns to Noah, carrying an olive branch in its beak. It may have been cultivation of the olive that allowed the nomadic pastoral tribes of Israel to settle down and build towns and cities. Israel supplied oil and trees to the Phoenicians, who in turn spread the tree to the Greek islands, probably around 1600 BC. The tree seems to have reached the Greek mainland between 1400 and 1200 BC.

In Mycenaean Crete, oil was produced from both wild and cultivated olives, with wild olive oil being preferred as a base for perfumes. The palace at Knossos had a huge oil store, reputed to be capable of holding 280,000 litres of oil, and single shipments of oil within Crete could be as large as 10,000 litres. By 1000 BC, there was regular trade between the Eastern and Western Mediterranean, and olive oil was its principal component. In classical Greece, the mythologising of the olive and its oil was well under way. According to legend, when Zeus was seeking a god to look after Attica, he held a contest to see who could come up with the best and most useful gift for humanity, the winner to get the job. Poseidon presented the horse, Athena the olive tree; Athena won. The story is told on one of the friezes on the Parthenon in Athens.

The first record of olive growing in Spain is from 1000 BC. By the sixth century BC, olive cultivation had spread to Tunisia, Tripoli and Sicily, and then moved north through Italy. The Romans, of course, were great promoters of the olive and huge consumers of its oil. They took olive cultivation with them on their conquest of the Mediterranean, and not just because they liked the oil. They could introduce order, provide the inhabitants with an income and guarantee Rome a good supply of olive oil. The Romans really got olive growing under way in Spain, and encouraged the planting of vast numbers of trees in North Africa and France. The oil they produced was traded throughout their vast empire, from Britain to Egypt, in ships filled with amphorae, the familiar traditional pottery jerrycans, each containing up to 100 kilograms of oil.

Olive growing continued to be important after the fall of the Romans, but production and trade became much more localised. The main centres of production through the Middle Ages were Spain and Southern Italy, especially around Palermo in Sicily. A later revival of olive growing further north saw new groves being planted in Tuscany and the South of France, and by the 13th century, Venice and Florence were established as very important trading centres for oil. The Visdomini di Tenaria, appointed in 14th century Venice, were probably one of the first regulatory authorities in the olive business, a sort of forerunner to today's International Olive Oil Commission (IOOC).

Much of the olive oil business since that time has been a family affair, with farms producing enough for their own needs and a surplus for sale in the market, where traders would buy and then move the oil around to where it was needed. It's really only since the arrival of mechanisation, over the last 150 years, and the creation of very large pressing facilities, that the business has recovered much of the international dimension that the Romans gave it.

AN OLIVE A DAY

Although it's not officially called the olive oil diet, a recent patent granted in the United States (No. 5 855 949) describes a detoxifying diet, in which patients are told to drink a slug of olive oil on the hour, every hour. This supposedly helps them to clear nasty pollutants out of their system, and lose weight at the same time – up to 5.5 kilograms a week if you also follow the diet's rules about eating only certain foods, and take an occasional top-up of desiccated thyroid gland to get the hormones moving. The olive oil's role, diet inventor Linsey McLean says, is to switch the body's mechanism over to burning fat, helping the body to get rid of the bad stuff that gets stored in body fat. It remains to be seen whether this diet will catch on with the Great American Public, the most overweight population in the world, but if it does, olive oil sales are going to boom.

A little less extreme, and a lot more popular, is the Zone diet propounded by American author Dr Barry Sears in a couple of best-selling books. It is otherwise known as a 40/30/30 diet, because it consists of eating food that contains carbohydrates, proteins and fats in that ratio at every meal. Sears recommends that virtually all the fat should be taken in the form of olive oil (or macadamia nuts – the only things with a higher content of monounsaturated fat). Not only does the olive oil give you energy, it also provides the precursors of several important hormones that Sears says play a critical part in health and general well-being. The Zone has become so popular in the United States that some restaurants have begun to sell 'Zone-favourable' dishes – and olive oil sales won't be doing badly either.

Even further into the mainstream, we have the Mediterranean diet, the currently trendy way of eating that features loads of olive oil and is said to

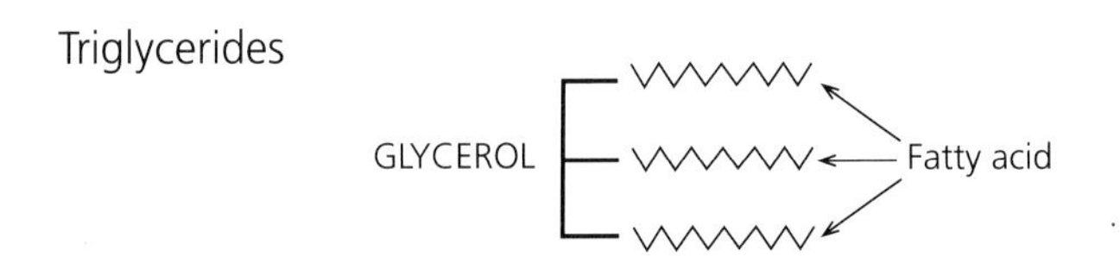

A triglyceride is three fatty acid molecules joined together by a glycerol molecule.

reduce both heart disease and the effects of ageing. Broadly speaking, it's here that modern medicine begins to catch up with ancient wisdom, putting olive oil at the heart of a healthy lifestyle.

At this point, the chemically illiterate may struggle a bit, but it's important to know something about the makeup of olive oil. Olive oil is 98.5 to 99.5 percent triglycerides – molecules made up of three fatty acids (long chain hydrocarbons with an acid bit on the end), linked together by another molecule called glycerol. Some of these fatty acids also float around without linking up with glycerol. These are the free fatty acids, which determine the overall acidity of the oil. The majority of the fatty acids in olive oil have 18 carbon atoms in their chain, and the single biggest component, oleic acid, has one double bond in its chain – making it a monounsaturated fatty acid. No double bonds and the fatty acid is saturated; more than one and it's polyunsaturated.

Oleic acid makes up between 55 and 83 percent of the oil. The next most common fatty acid is linoleic acid, at between 3.5 and 21 percent. It also has 18 carbon atoms, but two double bonds, and so is a polyunsaturate. The next acid in the sequence, linolenic acid, has three double bonds, but is usually present at levels of under 1 percent. Both of these last two acids are important because they supply the chemical precursors for a group of hormones called the prostaglandins. These are used by the body to control a huge variety of functions, including things like sex, inflammation, blood pressure and blood clotting.

The rest of the oil – the 0.5 to 1.5 percent that isn't triglycerides or free fatty acids – is made up of a mixture of pigments, antioxidants and volatile and aroma compounds. The pigments are principally chlorophyll, which is green, and carotene, which is red, and mixtures of the two give oil its characteristic range of colours. The antioxidants, mainly tocopherols and polyphenols, are very important, as we'll see, and include oleuropein, the bitter substance that gradually disappears as fruits ripen. The aroma compounds may be present in very small amounts, but add greatly to the flavour of the oil. Any refining process gets rid of these.

Scientific studies in the 1950s began to show that people who lived in the Mediterranean suffered lower levels of heart disease than those in northern Europe. The main difference between the two populations appeared to be diet. Northern Europeans, like the Scots, Finns and Dutch, all ate high levels of animal or saturated fats. Down south, they were eating lots of olive oil, whose principal fat is monounsaturated. Heart disease is closely linked to the level of cholesterol in the bloodstream, and that cholesterol comes in two forms: 'good' cholesterol, or high-density lipoprotein (HDL), and 'bad' cholesterol or low-density lipoprotein (LDL). Saturated fats raise the levels of both, while polyunsaturated fats can lower the levels of both. Only monounsaturated fats, such as the major constituent of olive oil, oleic acid, can keep the good HDL levels high, while lowering the levels of LDL, which is exactly what you need if you want to reduce your risk of a heart attack. In fact, Professor Don Beaven, a prominent New Zealand heart specialist, reckons that Australia and New Zealand could halve their deaths from coronaries if only the population could be persuaded to cut out animal fats (like butter) and take 30 percent of their calories in the form of olive oil.

So what is the Mediterranean diet? Aside from the oil content, it means eating like a Cretan peasant: lots of vegetables and fruit, plenty of fish and white meat, good wholemeal bread and other grains, and lashings of olive oil. It has the dual advantage of being good for you and tasting wonderful.

But if reducing your risk of death from heart disease isn't enough, olive oil has even more benefits. Good fresh olive oil is very rich in antioxidants, especially Vitamin E. The most active ingredient thereof, ∝-tocopherol,

Fatty acids

STEARIC ACID — no double bond — 'saturated'
CH_3 COOH (18:0)

OLEIC ACID — one double bond — 'monounsaturated'
CH_3 9 COOH (18:1 ω9)

LINOLEIC ACID — two double bonds — 'polyunsaturated'
CH_3 6 9 COOH (18:2 ω6)

α–LINOLENIC ACID — three double bonds — 'polyunsaturated'
CH_3 3 6 9 COOH (18:3 ω3)

All these fatty acids have 18 carbon atoms, but vary in the number of double bonds in the carbon backbone. Oleic acid is the principal component of olive oil.

works to protect the oil from oxidation, which is one reason why it can last for a long time without going rancid and also stand up to the rigours of frying food. In the body, the ∝-tocopherol fights the action of free radicals – little bits of highly reactive chemicals that can rampage through cells causing all sorts of damage. Free radicals are created perfectly normally in the body's day-to-day workings, but need to be controlled if they are not to cause too much damage, which can show up as ageing, liver diseases, rheumatoid arthritis, gout, psoriasis, emphysema, atherosclerosis, diabetes and cancer.

Another of the antioxidants that's beginning to make something of an impact is oleuropein, one of the polyphenols. Sold as olive leaf extract, it is now being touted as something of a wonder drug – nature's antibiotic, according to some, capable of dealing with viral and bacterial infections, including the common cold, herpes, arthritis, skin diseases, heart trouble of various kinds, diabetes and even AIDS. If it's as good as some of the sites on the Internet claim, then olive growers could well have another earner on their hands.

Olive oil is also very good for the skin. It may no longer be fashionable to smear your body with good oil before going shopping or running, but olive oil is widely used to soften dry skin and for massage. It's also traditionally used, if you really want to know, as a sexual lubricant.

OLIVES AROUND THE WORLD

Olive trees have been planted in just about every suitable climate zone in the world, and many more besides. As Europeans, especially the Spanish and Portuguese, explored and colonised the world, they took with them the tree that was so important to the economy at home. Even the British, not noted for the use of the olive in their domestic cuisine, found the time to trial olives in Australia, New Zealand and South Africa.

To better understand the global nature of the olive business, and how it has developed in the Southern Hemisphere, we need to take a quick tour of the world.

THE MEDITERRANEAN

Spain

Spain is easily the largest producer of olive oil in the world, a position it has probably held since Roman times. The main production areas are centred on Jaén, Cordoba, Toledo and Lérida, though olive trees grow in almost every corner of the country. Table olive production is a significant part of the olive economy, with most olives being processed green.

Spanish oil suffers from something of an image problem overseas, particularly in the world's largest import market, the United States. Canny Italian companies cornered that market years ago, and now a lot of Spanish oil is exported to Italy, where it is blended with other oils from around the Mediterranean and then sold on to the world market as Italian oil, or oil with an Italian label.

Spanish oil is, however, as good as any in the world, and some of its top producers, such as Nuñez de Prado from Baena, or the Oro de Genave

The Spanish olive-growing heartland between Jaén and Cordoba — virtually nothing but olive groves from horizon to horizon. *Olives Australia*

from the Sierra de Segura region in southern Spain, can stand comparison with the world's finest. Although it is impossible to generalise about the taste of oils from any one region, let alone the whole of Spain, it would be fair to say that they are often extremely fruity, and many are rich and nutty.

Most oil production is carried out by co-operatives or large companies, from mainly dryland groves of cultivars such as Picual, Arbequina, Cornicabra, Nevadillo, Blanquillo, Lechín de Sevilla and many others. Dual-purpose cultivars include Hojiblanca, Galego and others, while the principal table olive cultivars are Manzanilla (in many regional variations), Gordal and Picudo.

Italy

Italy usually finishes second to Spain in the world production table, but is easily the most influential. Only the alpine regions of Northern Italy fail to grow olives, but serious production occurs only as you move south across

A typical Tuscan hillside: olives growing on terraces, interspersed with other trees and other crops (frequently artichokes). *Olives Australia*

the Apennines into Tuscany, or west into the Ligurian hills. Tuscany, of course, has managed to establish itself in the world's eyes as the premier olive region, even if its production is small compared with the rest of the country. The warmer southern regions of Apulia, Calabria and Sicily are the real mass-production centres, accounting for 80 percent of the country's production.

When it comes to the marketing of olive oil, Italy is certainly the world leader. The Tuscan estates have almost single-handedly created a global market for expensive but tasty extra-virgin oil, where the label and single-estate status is (almost) more important than price. Many of the Tuscan aristocrats, such as Antinori, Coltibuono, where the original Chianti is said to have been made, and Frescobaldi, are also well-known winemakers and have brought that same expertise to the marketing of their oils.

Ligurian oils are generally said to be lighter, milder and sweeter than most other Italian oils, rarely having the strong pepperiness associated with Tuscany. The latter are generally held to be grassy and green, with a hint (sometimes more) of bitterness and lots of pepper, though much depends on the approach the producer chooses to follow – early harvest for the pungent stuff, later and riper for the fuller, fruitier flavours. Oils from the south tend to be richer and nuttier, with plenty of pepper.

The main oil cultivars (though there are hundreds noted in the literature) are Taggiasca in Liguria, the famous Tuscan trinity of Frantoio, Leccino and Pendolino, joined by Moraiolo, Maurino, Corregiolo, Razzaioi, Leccio del Corno, and many others, while Coratina dominates in the south. Table olive varieties include Uovo de Piccione, Nocellara del Belice, Tonda Iblea, Giaraffia and Ascolana Tenera, while Carolea, Moresca and Tonda di Cagliari are considered dual-purpose types.

Greece

Often second in the world oil-production table, Greece uses much of its product at home. Small growers dominate the business; they often have only a few trees producing fruit and the harvest is taken to the local co-op for processing.

Greek oil doesn't have the same sort of image as Italian products, and much of the best stuff is exported for inclusion in blends. Nevertheless, a few producers are trying to carve out a quality niche for their oils, and there's no doubt that these are of good quality. Table olives are also extremely important, especially the Kalamata olive and Greek-style cured olives – both regarded as among the best of table olives.

Most of Greece can grow olives, but Crete, the islands of the Aegean, including the famous groves of Lesbos, and the Pelopennese peninsula are important producers. Major oil cultivars include Koroneiki, Corfolia, Daphnoella and Daphonolia, while Kalamata is considered a dual-purpose olive, capable of producing a very good oil, as are Megaritiki and Vassiliki. Table olive cultivars include Konservolea, Halkidiki, Mastoides and Stravolia.

France

France is, in international terms, a very small olive producer, but its influence is much greater, partly because of the French people's love of and focus on quality food, and partly because the South of France has been an attractive 'home away from home' for many British and American expatriates.

Olive growing in France isn't easy. Periodic severe frosts destroy trees, and rapid changes in the rural economy have made small farms harder to maintain, leading to a steady decline in the numbers of olive trees being harvested. The rise of olive oil as a fashionable food has helped to slow down and perhaps reverse that process, and single-estate French oils are now widely regarded as being of the highest quality.

The growing regions stretch in a broad band along the Mediterranean coast from the Spanish to Italian border, and up the Rhône Valley as far as Valence. West of the Rhône delta, the main focus is on table olives, but in Provence, to the east, oil holds sway, except in Nyons, where the Tanche de Nyons is the only olive to be granted its own *appellation contrôlée*. Oil cultivars include Argental, Pendouiler, Pigalle and Rouget. Picholine, Bouteillan, Cailletier, Tanche, Salonenque and Grossane are considered dual-purpose, and Lucques is used for table olives only.

North Africa

Most of the oil produced by Algeria, Morocco, Tunisia and Libya is either consumed locally, or sold for blending on the international market. Table olives are important, especially in Morocco.

Most production is small-scale, from dryland trees that must cope with a very warm climate. Oil cultivars include Chemlali and Zalmati in Tunisia, Abelout, Faneya and Limli in Algeria; the Moroccans treat Manzanilla and Picholine Marocaine as dual-purpose cultivars. Table olive varieties include Meslala in Morocco, and Gerboui and Meski in Tunisia. Dual-purpose cultivars include Bouchouk Lafayette, Blanquette de Guelma and Sigoise in Algeria, and Barouni del Sahel, Limi, Marsalina, Tefahi and Yacouti in Tunisia.

Middle East

The largest producers in the Middle East are Syria and Turkey. Israel, a leader in the study of olive growing, is not a big producer. The industry there is intensive and highly mechanised, with irrigation wherever possible. In other parts of the region it tends to be much smaller scale and traditional in its approach to production.

In Israel, Barnea is probably the most important oil cultivar, with Nabali Mouhasan, Nabali Baladi and Souri being considered dual-purpose trees, and Kadesh, Manzanilla, Mehravia and Uovo de Piccione used for table olives. Syria has an oil cultivar called Zaity, and table olive trees called Abou-Salt, Djlat and Kaiss. Iraq's oil, — oil from trees, that is — comes from cultivars called Ajrosi, Barmaghi, Dikkam, Kasb and Jelin.

Other countries

In the rest of the Mediterranean only Cyprus produces a significant amount of oil. Olives are also cultivated all down the Adriatic coast of what used to be Yugoslavia, now mainly in Croatia. Olives are also grown in Iran and Pakistan, but I have found no information on their production. There is also a small but very lucrative olive-growing business in Japan, but organised more as a tourist attraction than an attempt to compete on world markets.

AUSTRALIA

The current boom in olive growing in Australia is far from being the first.

This tree, probably of French origin, is between 110 and 140 years old, and is one of 350 left of an original 1500 planted at Yakilo, South Australia.

The first olive arrived in Sydney at the end of December 1800, imported by market gardener George Suttor. He planted the tree in the governor's garden and brought more trees back from Europe in 1812. Western Australia received its first trees in 1831, and South Australia in 1836. Victoria seems to have waited until the 1870s and 1880s to see its first substantial plantings, but by that time interest in olives as a commercial crop was substantial, and planting was taking place in many areas. A depression during the 1890s, and the low price of imported oil, made it difficult for the new olive groves to make a living for their owners. During the early 20th century, despite sporadic attempts by various state governments to promote olive growing, there was little real development. What really turned things round was the arrival, after the Second World War, of immigrants from the Mediterranean, mainly Italians and Greeks, who were eventually to revolutionise Australians' view of food.

These immigrants were also amazed to find that there were plenty of olive trees growing around them, especially around Adelaide, and for many

Two icons and a tree — the new Australia.

An olive grove deep in the bush: Michael Burr's Beetaloo Olive Grove in South Australia.

years they were able to wander out into the countryside to collect olives for pickling, and for the small presses that a few people set up. Several large olive groves were planted in the post-war years, but all struggled in the face of competition from imports. During the 1960s and 1970s many of the surviving trees languished untended and unloved, but the food revolution of the 1980s, together with the worldwide growth in interest in fine olive oil, grabbed the attention of a new generation of potential olive growers, and the foundations for today's olive boom were laid.

While researching this book, I was amazed to discover the sheer extent of Australian interest in olive growing, and the commitment to establishing large economically viable groves. Everywhere I went, from Queensland to South Australia, people were talking about new projects and new presses. In the Hunter Valley, for instance, growers have established a co-operative to press their fruit as it becomes available, and are planning to install a large press in time for the year 2000 harvest. Co-op members already have between 150,000 and 200,000 trees in the ground, so their production is going to be substantial – certainly large enough to find a place alongside imported oils in the supermarket. There's also a flourishing market for boutique oils, made in relatively small quantities, but to very high standards, by a mixture of olive enthusiasts and winemakers. New groves are being planted in just about every state, from experiments in Alice Springs, to plans for hundreds of thousands of trees in Queensland. Large presses have been installed in Queensland and northern New South Wales. The spiritual heart of the business is probably still in South Australia, around Adelaide.

A common estimate of the number of trees likely to be in production by 2006 is 1.5 million, though others put the combination of recently planted and old established trees at nearer 3 million, all of which will be in production by the year 2000-2001. Oil production could be anywhere between 7000 and 9000 tonnes, and all of it should be very high quality. As Australia imports almost 18,000 tonnes of oil at the moment, the industry will have to take a substantial share of the local market if it is going to be viable. And the European Community (which includes Spain, Italy and Greece) subsidises its olive growers and oil presses and, until recently, directly subsidised exports. There is certainly a risk that, just as Australian growers begin to produce large volumes of oil, they will find themselves competing on a far from level playing field. On their side, though, is the fact that they are likely to be able to produce a better, fresher, tastier oil than the bulk blends being imported, and at a competitive price. It may not be cheaper, but it will certainly justify a premium if any of the oils I have tasted turn out to be typical.

Local growers' associations are becoming very active, and are a most useful source of information for anyone thinking of getting into the business. The Australian Olive Association is also extremely helpful, lobbying on behalf of the emerging business, and acting as a conduit for information. Contact details can be found in Chapter 15.

The original Australian plantings used trees imported from all over Europe, including Spain, France and Italy, and many trials were carried out to determine the best cultivars for the local climate and soil combinations. Unfortunately, as the business went through its boom-and-bust cycle, records were lost, trees torn out and seedlings went wild. The end result is that a huge variety of cultivars with many different origins are available for the intending grower. The most important are covered in Chapter 8.

NEW ZEALAND

Olive growing in New Zealand has a history almost as old as that in Australia. The first trees were planted by various settlers around the North Island, probably in the 1840s, and between 1860 and 1880 quite substantial groves were planted in the Auckland area. The remains of one of these can be seen today on Auckland's One Tree Hill. Nothing much happened for the next 100 years, until the government sponsored various trial plantings

A remnant of the early 1970s trial at Waipara Springs in North Canterbury — still fruiting happily after 25 untended years.

around the country during the 1960s and 1970s. These were successful, in that the trees grew well, but monitoring of fruit production was cursory and little attempt made to manage the trees for optimum production. Given that the average Kiwi of that day wouldn't have known an olive if it bit him, or used the oil for anything other than emergency tractor lubrication, it's perhaps not surprising that experimentation went no further.

The New Zealand business really got started in the mid-1980s, when Gidon Blumenfeld planted a grove in Blenheim. A friend of Israeli scientist and olive expert Shimon Lavee, Gidon imported Barnea, Uovo de Piccione, Manzanilla, Souri and Nabali Mouhasan with a view to propagating and selling trees. By 1990 he had 20 cultivars in his grove, and had imported New Zealand's first commercial oil press. Barnea was extensively promoted and planted widely around the country. In the late 1980s, the Blumenfelds were joined in their enthusiasm by artist Mike Ponder and his wife Diane, who planted olives alongside their grapes and laid the foundation for the now highly successful Ponder Estate olive oil. This is described by Judy Ridgway in *The Olive Oil Companion* as first-class, with 'a wonderfully strong aroma of freshly grated apple skins' and 'an interesting mixture of dried hay, bitter salad leaves, and a hint of lightly toasted almonds or chocolate' in the taste. In another significant development, Kiwi film-maker

Characteristically tall Barnea trees at the Muddy Water vineyard in Waipara, New Zealand.

Michael Seresin has established a grove of Tuscan varieties at his Seresin Estate winery in Marlborough, and plans to emulate the Ponders. There are probably more than 200,000 olive trees throughout the country.

Olive groves have now been planted in virtually all the suitable, and probably half-suitable, regions of New Zealand, from the sub-tropical north to the harsh continental climate of Central Otago. Most growers believe that, as with wine, the future of the New Zealand business lies in producing very high-quality oil for the international market. Local innovations include olive-growing subdivisions, where buyers looking for a new home site can buy into an olive grove. Several of these are under way in the Christchurch area and in North Canterbury around Waipara.

SOUTH AMERICA

In 1560, Spanish explorers took olive cuttings with them to Peru, and the tree followed the Spanish around as they conquered the continent. Olive growing became established in Chile, and to a much lesser extent in Brazil, but it is in Argentina that most modern development is taking place.

Over the last few years, as many as 1.2 million trees have been planted in the Catamarca region, to go with already substantial groves around Mendoza, San Juan and La Rioja. Driven by tax incentives, the business is gearing up in a big way, aiming to supply the large Brazilian market and then export to the world. Annoyed by European subsidies, they have imposed a tariff on imports to protect the local industry. With this level of commitment, there is no doubt that this rapidly expanding industry will soon have a significant impact in the world market.

Argentina's traditional olive varieties, Empeltre (a dual-purpose cultivar) and Arbequina, have been joined in the new plantings by Barnea, Manzanilla, Mission, Frantoio and Leccino.

UNITED STATES

California is the only part of the United States with a significant olive-growing industry. The olive was introduced by Franciscan missionaries soon after they founded their first mission at San Diego in 1769. Commercial planting didn't take place until the second half of the 19th century, and the first oil from these groves was pressed in 1871. The industry boomed between 1875 and 1905, when there were well over 500,000 trees in production, but the industry ran into competition from cheap imported oil, and soon switched over to table olive production.

California produces over 10 percent of the world's table olives, mainly black olives produced by processing green ripe Manzanilla, Sevillano and Mission fruit. The olives are mainly consumed in the United States, especially on pizzas. There has, however, been a recent upsurge of interest in oil production, led mainly by winemakers who have looked around the Napa and Sonoma Valleys and seen the legacy of fine old trees left behind by the 19th-century settlers. Old Picholines from France are there, along with Spanish and Italian cultivars, and a great deal of fine oil is now being pressed, aimed at a resolutely upmarket niche.

SOUTH AFRICA

Olives arrived in South Africa in 1661, but didn't really become an established crop until the late 19th century, when Jan Minnaar planted trees in the Western Cape. These were only experimental plantings, but did well enough to win him a Gold Medal at the 1907 London Show for the finest oil produced in the British Empire. In 1925 an ambitious Italian nurseryman started planting olives, again in the Western Cape, and by 1935 was installing a press. By the late 1960s, around 100,000 trees were in production, with 70 percent of them around Paarl.

The principal varieties in use include Leccino for oil, and Mission and Manzanilla for table olives. There are rumours of substantial plantings taking place at the moment, and it may be that another player is about to enter the world olive scene.

FROM TREE TO TABLE

Olives ripen as winter sets in. Growers who live in 'winterless' areas like Queensland or Southern California may not freeze like the rest of us, but they also may not be able to make the superbly flavoured oil to which we colder-climate growers aspire. Whatever, it will always be winter when you harvest. In the Southern Hemisphere, the harvest should be over by August, though in South Australia harvesting and pressing can run on into September.

The romance of the traditional olive harvest has been dealt with by many authors. Frances Mayes's best-selling *Under the Tuscan Sun* is full of evocative detail about her experiences in rejuvenating some old olive trees, picking the fruit and making good oil. And Mort Rosenblum's *Olives* is a treasure trove of olive harvest lore (and just about anything else to do with the trees). It's very easy to be seduced by the glowing prose, even if it is tempered by sobering descriptions of the hardships and difficulties.

Back in the real world, the aim is to pick the fruit when the delicate balance between oil content and flavour is just right: too soon, and the oil will be green and peppery – perhaps too bitter to use – and the yield will be low; too late, and it will be flat, bland and have a higher level of acidity.

Perfect ripeness depends on the tree, the season and the climate. What's right for Moraiolo in Tuscany may be wrong for the Moraiolo in my front paddock, and certainly won't apply to Picholine in Provence (or Paraparaumu). As we saw earlier, olives can be green ripe as well as fully black ripe. The rule of thumb used by Tuscan peasants (and aristocrats) since time immemorial is to pick when half the fruit is turning from reddish-green to black. The proportion of green fruit gives Tuscan oil its grassiness and pepperiness. In parts of Greece, nets are spread under the trees and the olives allowed to ripen on the tree until they fall off by themselves. Not a recipe for great oil, but good enough when labour is scarce. Shimon Lavee recommends that to get the best balance between oil yield and flavour, you should pick when the fruit has just turned completely black – but his taste may differ from yours.

Tradition, saints' days, phases of the moon and impending frosts all play their part in determining the harvest date in traditional olive regions. Another crucial factor may be the availability of a time slot at an olive press you deem suitable (old-fashioned and romantic, stainless steel and modern, or with an honest owner who won't rip you off).

So harvest day dawns, you gather your (preferably extended) family around you and head off to the grove. You will have ladders, baskets, nets, little plastic hand rakes (or goats' horns if you're Tunisian) and maybe some machinery if you're modern. The aim is to get the fruit off the trees and into the press as quickly as possible. The longer the fruit moulders in storage after picking, the higher the acidity of the oil and the lower its value. If the fruit is clean and unbruised, so much the better. If you're picking your fruit for pickling, you'll be especially careful to avoid bruising.

One traditional European picking method is to spread the nets under the trees, and then to wander through the grove systematically banging branches with sticks so that the ripe fruit falls off. This is effective with the right cultivars and if the fruit is ripe enough, but it can damage the tree and reduce the following year's crop, so it's probably best avoided. A less drastic, but more precarious approach is to hand pick, using rakes or your fingers, climbing high inside the tree or around the outside on ladders. The approved method is to 'milk' the branches, running your hand along towards the tip, raking off the olives with your fingers. You can pick into wicker baskets slung round your waist, or just let the fruit fall into the nets spread on the ground, trying to avoid breaking off too many twigs and leaves in the process.

Once off the tree, the olives are put into sacks (traditional) or plastic crates (modern) and stacked ready for transport to the mill. This should be a cart drawn by a donkey, but these days small tractors are more common.

The fruit is first washed to remove any dirt and leaves (though some Tuscan growers might leave a few leaves in to 'improve' the flavour of the oil), and then fed into a crusher. The most picturesque form of crusher involves large granite millstones rotating over a steel tray, crushing the

Olive paste being squeezed onto modern nylon pressing mats. *Helen Clausen*

whole fruit, seed, flesh and all, into a paste. The modern alternative is a hammer mill, which chops up the fruit with rotating knives. Either way, the paste then has to be kneaded for up to an hour to allow the oil to coalesce into droplets. If the paste is too thick, a little water may added. Heating the water increases the yield of oil, but may also leach out some of the flavour – hence the term cold-pressed olive oil, though in some places 'cold' may be interpreted to mean water as hot as 30°C.

The paste is then squeezed out like toothpaste onto mats – traditionally made from esparto grass, but now more usually nylon – and stacked under a large hydraulic press that can exert tons of pressure. As the pressure increases, so the oil flows out at the edges of the mats and drops into a collection tray. Unfortunately, more than just oil comes out: a great deal of olive juice – a dark black, bitter water – is produced, and has to be separated from the oil. If the oil and water mixture is allowed to stand for long enough the oil will float to the surface and can either be skimmed off, or the water drawn off from a tap at the bottom of the tank.

A modern oil press uses large centrifuges to spin the oil out of the paste, and can separate the oil from the black water in the process. These machines, which can be very large indeed, can operate continuously, processing many tons per hour – an essential feature if large amounts of oil are to be produced at an economic cost. A typical small hydraulic press might hold 100 kilograms of paste and take an hour to get all the oil out. The mats then have to be reloaded with the next batch, so the fruit from even a small grove will take many hours to press.

There's a certain amount of controversy among purists about the use of these giant centrifuges. Some maintain that they produce less tasty oil, that it is somehow brutalised by the process; for them, only oil produced in hydraulic presses can be really good. Others will tell you that this is nonsense – many of the world's finest oils are centrifuged in modern pressing facilities. Exceedingly fine oil can be produced by both methods.

It is also worth pointing out that some experts reckon that the best oil can come only from the 'flowers' of the pressing – that is, the oil that runs freely from the paste before any pressure (other than its own weight) is brought to bear. You can buy the stuff, and very expensive it is, too. Others, of a more biblical bent, may tell you that the finest oil is 'washed' oil, extracted just as it was in the 1000 years before Christ, by crushing the fruit and then soaking it in water, finally skimming off the oil that floats to the surface. This is almost certainly uneconomic in today's terms.

When it leaves the press, the new oil is often filtered to remove any cloudiness caused by stray bits of fruit. On the other hand, many producers of the best oils leave the fruit haze in the oil, believing that it adds to the flavour. Unfiltered oil will also clear naturally over time, throwing a deposit on the bottom of the container. This could shorten the shelf-life of the oil, so it is safe to assume that the oil you buy in your supermarket has been filtered, unless the bottle specifically tells you otherwise.

After pressing, the acidity of the oil will be tested. This will determine whether you can call it extra virgin, virgin and so on. Below 1 percent free acid, you can call it extra virgin, and that's what most modern growers aspire to produce. It's the best oil, and it commands a premium price. And you will taste it; on a piece of bread, if you're in Tuscany – *il prove del pane* they call it – or in a little plastic tasting cup, or on a teaspoon. It could even become an obsession. The moment when you taste the first oil made from your grove is magical.

If you're a small olive grower, your harvest may end there, in a few tens or hundreds of litres for your own use or for sale from your garage. It might also have ended earlier, at the gate of the press, when you sold your fruit direct to the press or to the co-operative. In that case you'll be waiting anxiously to hear what percentage oil your fruit produced, at what level of acidity, and so what you'll be paid.

The larger grower, or the olive oil estate, might harvest in several stages as various varieties ripen at different times – harvesting some earlier and greener for extra flavour, others later and mellower to fill out the palate of the final blend. Then the complexities are limited only by your trees, the season and your imagination. Even small growers may aspire to something similar, experimenting with their fruit until they arrive at an approach that gives them the kind of oil they're after.

Oil storage is relatively straightforward. It should be kept fairly cool, and away from light. Fresh olive oil can last for up to a year, perhaps slightly more, before it begins to go rancid. At the same time, new oil may need a few months in the bottle if its combination of flavours is going to settle down and harmonise. In the olive heartland the new season's olive oil is eagerly awaited, and there are signs that this is beginning to spread to restaurants and food enthusiasts in the rest of Europe and the United States. Just as each new season's wine releases from major labels are viewed as important milestones, so a few restaurateurs and retailers are beginning to make mileage out of their search for the best of the new season's oil. Rose Gray and Ruth Rogers of London's trendsetting River Café are keen exponents: 'New oil poured over hot bruschetta is a gastronomic experience we look forward to every year, for now that our reputation has spread, we are invited to Italy to taste the new oils.' Lucky people.

Olives for pickling face a very different fate. Far from being crushed to paste and squeezed till their pips squeak, they will be mollycoddled and carefully preserved, before being finished in exotic combinations of herbs and oil. Table olives have to be picked very carefully, almost invariably by hand, because any bruising will quickly spoil the look and flavour of the finished product.

Bite a ripe olive and your mouth will pucker up as the intense bitterness hits your tongue. Caused by the presence of oleuropein, this bitterness must be removed before the fruit can be considered edible. If you leave the fruit on the tree long enough, the bitterness will disappear of its own accord, but you will then have to fight the birds for the fruit and, because it will have been on the tree throughout winter, flowering and fruit set in the following summer will be reduced. There is, however, a Greek variety called Throubolea, grown in Crete, the Aegean islands and parts of the mainland, which loses its bitterness more quickly, and which requires little or no treatment before sale. It is said to be popular with people suffering from heart and kidney trouble.

There are two main methods of getting the bitterness out of olives: soaking them for a long time in salty water, or treating them with lye, which is a solution of caustic soda or sodium hydroxide. In either case, the objective is get the oleuropein out of the flesh of the fruit, while leaving behind as much of the flavour and goodness as possible. A third method is to take very ripe olives and dry-cure them in salt, a Greek speciality. Between those three methods lie a myriad of different techniques, varying from region to region, traditional and modern, to local delicacies found only in parts of Sicily or Morocco.

Pickling olives is not a quick process – it can take several months to create the final product, and there are many pitfalls along the way. Some of these are potentially fatal to the end user, so it's perhaps a process best left to food processing professionals. Nevertheless, the basic process is simple, and with care a grower can easily pickle enough for his or her own table. A selection of recipes to produce different styles is given on pp130-2.

TASTING AND USING OLIVE OIL

If all you've ever experienced in life is olive oil from a chemist, or cheap oil from a supermarket, then your first taste of the real thing may come as something of a shock. Treat yourself. Buy a bottle of the most expensive extra virgin olive oil you can find. It'll probably be in a rather posh package – a dark glass vessel with either a metal screw-top or a cork, perhaps even a wine-bottle cork that needs extracting with a corkscrew. Read the label, especially if there's one on the back. It should tell you where the oil comes from, may tell you what varieties were pressed, the acidity and could even include a genealogy of the count who owns the trees, or some other romantic marketing 'story'.

Open the bottle. Look down the neck at the liquid and take a deep sniff. Then pour a little onto a piece of good bread, or dip the bread into a small puddle of the oil on a white dish. Look at the colour of the oil – it could be grassy green through to a rich yellow, almost amber. Finally, and with due reverence, taste the oil. It will certainly taste of olives, possible more olivey than you thought possible, rich, redolent of the fruit, intense, even exciting if you're of a gastronomic turn of mind. Taste some more, then make a simple salad dressing with the stuff: 90 percent oil, 10 percent of the best white wine vinegar you can find, a little salt and freshly ground pepper. Toss a few fresh salad leaves in the dressing, then eat.

Using the best olive oils has a lot in common with drinking fine wines, from the anticipation, through the appreciation of the aroma, to the tracking of the flavours as they cross the palate. And then, of course, there's the price. Top olive oils can easily cost more per bottle than the best wines from the same countries.

As in the world of wine, all olive oils are not equal. Different varieties of olives produce different flavours, climate and *terroir* have their influence on the end product, and crude commercial reality intrudes at many points. Oil companies produce many different brands and grades of olive oil, designed to meet varying quality and price objectives. They blend oils from all over the Mediterranean to achieve the styles they want. As a result, your local supermarket shelf will certainly have a range of different grades of oil, in varying sizes and at varying prices.

The International Olive Oil Council (IOOC) sets the commercial standards for the classification of olive oils, using a range of simple and complex chemical tests, as well as tasting and rules about the method of oil production. The full definitions are complex and demand some knowledge of oil chemistry, so I'll leave them to the experts (Michael Burr's *Australian Olives* covers them very well), but whether you're a grower or a shopper, you need to know the broad definitions.

Anything calling itself olive oil has to be made exclusively from oil extracted from olives. No other oils can be added. That may sound like an obvious enough sort of rule, but you should bear in mind that olive oil is expensive compared with other vegetable oils, and good money can be made by 'cutting' olive oil with something cheap and innocuous.

Virgin olive oils are produced solely by pressing or centrifugation, without using chemical extraction or excessive heating. They are graded into four levels, by looking at the level of free acidity in the oil, and by sensory or organoleptic rating of the taste. The latter process is mainly for spotting defects rather attributes, and can be done only by accredited tasters. If they are not available, then acidity alone will do.

Extra virgin olive oil must have an acidity of less than 1 percent (1 gram of free acid per 100 grams of oil), taste "fruity" and have no taste defects.

Virgin olive oil has an acidity of between 1 and 2 percent, should be fruity and have defects that rank under 2.5 on the IOOC's organoleptic rating scale.

Ordinary virgin olive oil has an acidity of between 2 and 3.3 percent, and if it is fruity, may have other defects up to 6 on the IOOC scale, or, if not fruity, fewer than 2.5 defects.

Any virgin oil that has more than 3.3 percent acidity, or scores more than six on the defects scale – which means that it tastes pretty awful – is automatically deemed unfit for human consumption unless purified. Such oil is called lampante, a reference to the importance of olive oil for burning in lamps.

Refined olive oil starts out as virgin oil, and is then treated chemically – presumably to reduce acidity or remove off tastes – without altering the chemical make-up of the oil (the glyceride structure, to be precise).

Olive oil (at last, plain ordinary olive oil) is a blend of refined olive oil with some virgin olive oil (but not, obviously, lampante) to give it back some flavour.

Crude olive-pomace oils are produced by treating the pomace – what's left of the olive skin, flesh and stone after the oil has been squeezed out – to extract the remaining oil (usually around 5-8 percent of the total weight). The pulp is first dried, then treated with solvents to extract the oil.

Refined olive-pomace oil is made from crude olive-pomace oil, leaving the glyceride structure unchanged, and can be sold for human consumption either alone or blended with some virgin oil for flavour.

Olive-pomace oil is a blend of refined olive-pomace oil and some virgin oil.

You won't see all of these types of oil on the supermarket shelf. There will be various brands of extra virgin, virgin and olive oil, and there may be tins of oil that refer to olive-pomace oil on the label but, outside of the gourmet section, the oils you see will be mass-market stuff, produced and blended in factories, and often with only the slightest traces of the original flavours left in them. As an intending grower, you should seek out the best oils if you want to find out what the business is really about.

Tasting olive oil for professional purposes – organoleptic assessment, as the scientists call it – is a mixture of science and art. If you've ever been befuddled by wine connoisseurs and their language, then be warned: the same thing happens in the olive oil business. But just as wine professionals have systematised their use of language, so olive experts have been developing their own language and agreed parameters for the subjective assessment of oil quality.

The IOOC system used to define oil quality first seeks to establish the presence (or preferably absence) of six main defects: fustiness, mustiness, muddiness, winey-sour-acid-vinegary, metallic and rancidity. There is also a category for 'other defects', which includes heated or burnt, hay/wood, rough, greasy, vegetable water, brine, esparto, earthy, grubby and cucumber, but these are apparently not yet widely used. Each term is related to a particular flavour in the oil, and it is possible to learn how to recognise

Ready to taste: oil, white plates to show off the colour, and lots of good bread — in Tuscany it would be salt-free bread. *IOOC*

these characteristics, preferably under expert guidance.

Of the six main defects, rancidity, the nasty flavour produced as the oil is oxidised – something that happens naturally as oil ages – is the most damaging. Fusty is the taste produced by olives that were stored for too long before pressing and began to ferment. Musty comes from fungi and yeasts growing on olives that were stored in high humidity, while muddy sediment results from keeping the oil for too long on sediments that form naturally in unfiltered oil. Metallic flavours are just that, caused by contact with metals during processing, and winey-sour-acid-vinegary results from the production of acetic acid and other chemicals in the oil. Too much of any of these defects, and the oil is lampante.

If the oil survives that stage, then the tasters go on to look for three positive attributes: a fresh olive fruitiness, which can be ripe or green, plus bitterness and pungency. Fruitiness is self-explanatory and readily perceived, but bitterness and pungency are less so. Pungency means a sort of intense biting sensation in the mouth, and later the throat. It is not a difficult thing to spot, but can be an overwhelming experience and last for a fair while. Bitterness, technically, is a passing sensation on the tongue owing to some greenness in the fruit. Most importantly, these three characteristics have to be present in moderation and in balance.

The actual tasting of oil isn't an overly complex process. It helps if you don't smoke and haven't just had squid in a chilli sauce for lunch. The oil should be presented in a glass or little plastic tasting cup. Put your palm over the top of the glass to keep the aroma in, and then cup the bowl in the other hand to gently warm the oil and to help it to release its aromas. When the oil has warmed up, swirl it around a bit, and take a first sniff, putting your hand back over the glass. Then take another, deeper snort. Finally, sip a little of the oil, curl up your tongue so that you've got a little puddle of oil behind your front teeth, and then suck air in over the top of the oil so that it sprays over the inside of the mouth.

I don't pretend to be an expert at tasting oil, but I do know that if I'm even to pretend to make fine oil from my little grove, then I'll have to train myself and my palate. And it is certainly a process that can be learned, and that is repeatable. I once presented an IOOC-accredited taster with three oils to try. She pronounced on all three, very succinctly and definitively. I, an almost complete virgin to the tasting trade, could see something of what she was on about, but when I presented the same three oils to another three IOOC-accredited tasters the following morning, they came back with almost identical opinions on the strengths and faults of the three oils.

Of course, oil tasting would be a pretty dull affair if all we did was look for defects and a few simple positives; there has to be more to life than pungency – and there is. Australian olive expert Michael Burr, himself an IOOC-accredited taster, lists among others, such flavours as apple, almond, grass, pressing mat, green leaves, and harsh. Some are positive, some negative. Pressing mat, for instance, is a taste produced by using pressing mats that haven't been properly cleaned or stored. Bits of the olive paste cling to the mat between pressing sessions, and if it's warm or too long elapses between pressings, it can begin to ferment or go off, tainting the next batch pressed.

Other researchers have used (and presented in an assessment wheel) tastes such as putty, caramel, tomato, chilli, green banana, orange, fruit candies and wild flowers. The sky's the limit, it seems.

Back in the real world of the kitchen, and in choosing oils to use in everyday cooking, there can be no gainsaying your own personal preferences. If it's what you like, then that's okay – as long as you've taken the trouble do a little research first. Most people wouldn't go so far as Mort Rosenblum, who happily admits to having at least a dozen oils at any one time, each one matched to a particular purpose – and not all of them extra virgin. Others might find it a tad pretentious to have five or six different single-estate extra virgin oils for different dishes or cooking methods, but there really is method in this madness. An oil that's superb on salad may be too good (or expensive) to use for frying potatoes. Another may bring steamed vegetables to life, but not cut it when dressing pasta. A third might be fiery and pungent tasted from the bottle, but work really well when used in browning chicken or beef, or starting a bolognese sauce.

The simplest approach, and the least controversial, is to look for a good mass-market oil for general frying purposes (though deep-frying in a suitable extra virgin oil can produce stunning results), a good mass-market extra virgin for dishes that need more finesse, and one or two (or as many as you can afford) really good single-estate cold-pressed extra virgin oils with different characters for use on salads, steamed vegetables or in any of the classic olive oil dishes (see Chapter 14). Most olive oils in any category will not survive for more than one year, so don't buy huge quantities of oil at any one time, and keep what you do have in a cool, dark place. Too much light and heat can speed up the natural ageing of the oil, reducing flavours and perhaps inducing rancidity – but don't put the stuff in the fridge. Most olive oils will go thick and waxy, throwing a white precipitate at the bottom of the bottle if stored below about 10°C. This doesn't hurt the oil – warming it up will get rid of the white stuff – but it does make it hard to use straight from the fridge!

I suspect that this sort of 'standard' advice will have to change over the next five to 10 years. As new growers and co-operatives, particularly in Australia, begin to produce large volumes of extra virgin oil, buying oil by IOOC grade may become irrelevant. You'll want a mild, fruity extra virgin for frying, something a little more assertive for general cooking, and a few single-estate extra virgins for salads and the like. The differentiation will be on flavour and purpose, not free fatty acid content, because new growers will want only the best, at the best possible price. And that's a prospect I find profoundly exciting.

GETTING STARTED

When my wife and I first decided that it would be 'nice to have a few olives on the front paddock', I had no idea what realising that particular little dream would involve, nor how lucky we were in a number of very important respects. We were lucky with the land – a north-facing bowl sheltered from cold southerly and easterly winds, just about frost-free, fertile and free-draining, and not as exposed to the searing Canterbury nor'wester as the rest of the farm. We were lucky that it was big enough to support somewhere between 200 and 300 trees, just about the minimum number required for a commercial venture, and fortunate that we also had other income-generating tree crops in the ground, to reduce risk. We were lucky with the climate – about as Mediterranean as it comes in New Zealand, with hot, dry summers and mild wet winters, and blessed by being in one of the fastest-growing wine regions in the country. And we were lucky because we were coming to the business from, for us, the 'right end' – a basic comprehension of the real standing of olive oil and olives in fine food. So when we began to think about our olive grove, we naturally set about trying to re-create something of the fine oils we had experienced in our years in Europe.

Our good fortune was a matter of serendipity, or happy chance. I cannot recommend that other olive growers should act on the same basis. You need to make a series of well-thought-out, educated decisions long before you start planting trees. And if you hit a problem early, you should also be prepared to change your plans radically – or perhaps even give them up altogether.

Initial planning

There are a host of reasons why people consider planting commercial or semi-commercial olive groves. On one level, there's the agribusiness investor with plenty of capital to invest in a crop that has the potential to generate a good medium-term return, and perhaps also has tax advantages. At the other end of the spectrum, you have the so-called lifestylers, with their smallish blocks of land and a sentimental attachment to the idea of olives as a symbol of health and wealth. In between there are a myriad possibilities. There are farmers looking for a productive use for second-grade land, or perhaps to diversify away from pastoralism into something that can yield a higher return per hectare, or simply to find a better use for land that for years has been growing grass when it could have been much better used growing trees. Property developers may want to find crops that can meet economic use criteria when subdividing small blocks out of larger farms. Others may be looking for something that can generate a retirement income.

Whatever scale of operation you have in mind, you must have an idea of what you're going to do with the crop when it finally arrives. The olives off a couple of trees in the back garden or home orchard can be pickled for family use, perhaps even pressed for oil if you know someone with a small press, but when you're planting hundreds of trees that will, in a few years, produce tonnes of olives, some sensible forethought is clearly required. You also need to factor in how dependent you'll be on the income generated by your trees. If the olives are only a sideline, then you may be prepared to accept a level of risk that might scare someone else rigid. And if you're going to rely on the olives for most, or even all, of your income, then you will need to be very conservative in your planning.

Establishing the level of risk in planting olives on your property, or selecting a suitable piece of land to buy, demands a careful appreciation of the interaction between climate, soil, site position and layout, the cultivars you plant, and the way you plant and then manage them. Depending on the levels of risk you're prepared to accept, you may find that you're happy to proceed, or you may need to look for another crop altogether.

Climate

Olives have a reputation as tough trees, able to withstand drought, heat and, to a lesser extent, cold. Surviving, however, is a very different thing to thriving. Although big old trees may manage quite happily without irrigation on the edges of the North African desert, or cling to sun-bleached white cliffs on a Greek island, they are almost certainly uneconomic in any terms that a Southern Hemisphere grower might understand. To be a profitable fruit producer, the tree needs to be pampered. A little stress may be a good thing – as it is with grapes, helping to produce flavours in the fruit – but, if taken to extremes, it will limit tree growth and fruit production, and make the economics of olive growing rather unattractive.

The perfect climate for olives is hard to define. Old established trees may be able to take occasional exposure to very low temperatures when dormant, but 24 hours or longer at -8°C will damage most olives, and the younger they are, the more damage they'll suffer. Young trees direct from the nursery may struggle after an air frost of -3° to -4°C, with growing tips showing the worst effects. Just about all the northern Mediterranean's olive-growing regions, with the exception of parts of Spain, southern Italy and those parts of Greece near the sea, can experience weather cold enough to kill olive trees. Particularly damaging are cold snaps towards the end of winter which follow weather mild enough to convince the trees to start spring growth.

Frosts may also be a problem if they occur in late spring, damaging flowers and thereby reducing fruit set, or early in autumn, freezing the fruit on the tree before it can be harvested. If your proposed site is known to suffer regularly from this sort of frost, even though winters are generally mild, it may be enough to put the economics of the business in doubt.

The best advice for growers in colder areas is to select sites that have a good aspect to the sun, to avoid known frost hollows and to plant varieties known to be cold tolerant. Tuscan varieties in general, and Frantoio in particular, are said to stand frost well, and a couple of New Zealand nurseries claim that Nabali Mahousan, SA Verdale, Picholine, California Mission, Hojiblanca and Picual are cold tolerant. It's also possible to manage the grove and trees to reduce the impact of frost; we'll look at that later.

The paradox, of course, is that the olive requires a period of winter chilling to help initiate flowering. If the tree doesn't get enough cold (and the definition of 'enough' is pretty broad, from 1200 hours per annum below 7°C to perhaps as little as 50 hours), then flowering will be reduced and fruit yield lowered. This requirement varies from cultivar to cultivar, and there are certainly plenty of trees doing well enough in North Africa (and Queensland). Nevertheless, growers in regions with mild to warm winters should probably use varieties from warmer areas of the

Mediterranean, or to look to growers around them for what's doing well. In Australia, Olives Australia, the huge nursery in Queensland, is researching the topic, and will certainly be pleased to help.

So the olives need a coolish winter without damaging frosts, and a mild spring. Summer heat is also obviously required, and a relatively long, mild autumn to allow the fruit to ripen. But there's a paradox here, too. If

This poor little tree is struggling in ground made wet by a sub-surface spring. Compare it with the healthier trees behind.

the temperature rises above 33°C for more than two to three days, the tree will shut down and become dormant, waiting for temperatures to moderate. In regions where this is common, such as North Africa, the Middle East, parts of Spain and chunks of Australia, olives do most of their growing in spring and autumn. Irrigation tends to reduce this effect, and also maximises the growth spurts.

Water

Olives hate having wet feet. Trees that have too much water around their roots for too long will not thrive, and may die. Episodes of prolonged flooding, or regular winter waterlogging, would certainly be too much for olive trees to take. But the precise relationship between annual average rainfall and the viability of olive trees depends on a number of factors, including the type of soil and its ability to hold water, whether the grove is on a slope or particularly free-draining sub-soils and the availability of water for irrigation. One rule of thumb, quoted by Mike Ponder in *The Good Oil*, is that if you think that your olives won't need irrigation, then you're probably in an area that's too wet to grow olives.

Various approaches can be used to make damp sites more appealing to olives, including deep ripping to improve drainage, or even installing a drain network. It might also be possible to plant the trees in raised mounds of suitable soil, to mitigate the effects of flooding or soils that hold too much moisture. If you're having to consider these options, however, you may well be better off looking for another site.

Areas of high rainfall often also suffer from high humidity, especially as one gets closer to the equator. This may also limit the vigour of olive trees, because humidity encourages fungal growth, and most cultivars are susceptible to attack by various kinds of fungi (see Chapter 11). At the very least, the trees will require more spraying than those growing in a drier climate.

Olives are grown without irrigation in areas with as little as 200 millimetres of rainfall per annum, but fruit yields are unlikely to be very exciting. In most traditional olive-growing areas, annual rainfall averages between 400 and 700 millimetres, and in commercial groves in these areas, at least occasional irrigation would be necessary. All the research shows that irrigation increases fruit yield and can reduce biennial bearing, so the availability of water is a crucial factor when considering potential sites.

How much water you'll require is another of those 'how long is a piece of string' questions. It will vary depending on the size of your grove, the natural rainfall that season, the soil type and its holding capacity, whether you have grass between the rows or bare earth, how much wind has been blowing and the method of irrigation you choose to use. You should try to discover your local 'evapotranspiration'(ETP) rate (often published in local papers). This is the amount of water lost on a daily basis through a combination of evaporation from the soil surface, and transpiration by the plants growing in it, expressed as the equivalent amount of rainfall in millimetres. In my part of New Zealand, the ETP on a normal summer day will be between 3 and 4 millimetres, but if there's a nor'-wester blowing, it could rise to over 12 millimetres.

In general, irrigation is essential to help young trees survive the transition from pampered nursery to rough old paddock, and to ensure that the trees grow well in their early years. Once they are established and productive, it may be possible to reduce the amount of water you apply, and to manage the timing of the applications to influence factors such as flesh-to-pit ratio, flower set and fruit size.

Olives are moderately tolerant of saline irrigation water, which is often a problem in Australia. But keeping the soil – and trees – healthy when using saline irrigation water can be a problem, and you should consult local experts before making any rash decisions.

Wind

The exposure of your grove to wind is also a factor to consider, whether you suffer from hot dry winds such as my beloved nor'wester, or cold, dry winds such as the mistral of the Rhone Valley and Provence. Even a plain old wind, if sufficiently steady and prolonged, can have a distinct impact on tree growth, imparting a lean to the trunk. Some varieties are more prone to this than others, and Leccino in particular is reported to be the worst. Others, such as Koroneiki, are said to be wind-resistant, so it would be sensible to put your Leccino in the most sheltered part of your site, and your Koroneiki on the windward side of the grove. Wind damage is easy to spot, and I don't mean bits of tree all over the grove, though that happens. A hot, dry wind blowing to gale force can suck water out of leaves faster than the trunk can pump it out of the ground. Olives are better adapted than most trees to resist this effect, but they will still suffer if exposure is prolonged. The dehydrating effect may be particularly severe on flowers.

In the absence of established shelter, young trees can be protected from

Cold air is heavier than warm air, and flows down slopes. Any obstruction will tend to make it collect, and create a frost hollow or pond.

at least some of the damaging effects of wind by using tree guards of various kinds. These also provide some frost protection, and certainly keep rabbits and hares out. Good stakes are also essential.

Using olives as shelter may seem profligate but, if given water, and planted about 2 metres apart, they will grow into a tall, vigorous hedge. They will also help to increase the amount of pollen flying through the grove during flowering, which is bound to be helpful. If you have to plant shelter, and want to use more traditional species such as poplar or macrocarpa, you will need to ensure that the new trees don't spread their root systems into the olive grove (by allowing a bigger gap than you ever thought necessary), or annoy you by suckering all over the place. This will be more of a problem if your soils are shallow. You should also be careful that your new shelter belt doesn't act as a cold air trap, increasing the risk of frost damage to the trees. This is because cold air has a nasty habit of flowing down slopes and will collect in front of any obstruction, such as a shelter belt or a wall, that impedes its progress. If that area is also sheltered from the sun during winter months, frost may linger there all day, when elsewhere the traces of a ground frost vanish soon after sunrise. Given a run of clear nights, the cumulative effect can be striking. There's a corner of my garden like that, and after a few clear days in winter the soil can be frozen up to an inch or more deep, while a couple of metres away the citrus trees are looking perky and happy with their lot.

Soil

What you see at the surface doesn't tell the whole story. Soil changes with depth, and there may be layers of different soils and sub-soils before you get down to bedrock. In places where plains have been built up by glacial and river deposits, there may be hundreds, or even thousands, of metres of gravels and clays before bedrock is reached.

The standard advice is to get to know the soil of your potential olive grove, and its profile, and, if necessary, to get professional help from a consultant. It can be a complicated business, and expert advice will help to get the trees off to a good start. The first step is to arrange for a soil analysis. This might cost a few dollars, but it will be money well spent. It will tell you what's going on in the soil, giving you a picture of its acidity or alkalinity and the levels of various key nutrients that will be available to the plants. Don't just take a sample from one point on the paddock – take several from various locations so that you can get an idea of average conditions. You should also be sampling down in the region where the tree will have the majority of its roots – perhaps around 30 to 70 centimetres below the surface.

Olives are generally regarded as shallow rooting, and in the shallow soils typical in traditional olive-growing areas that's certainly true. Most of the root system is found in the top 70 centimetres of the soil, with more down to 1.2 metres. In deeper soils, olives have been found to send roots down 6 metres or more and, in stratified (layered) soils, to develop stratified root systems. Irrigation tends to keep the roots shallow.

Olives will tolerate a wide range of soils, from the moderately acid (pH6.5) to the moderately alkaline (pH7.5 to 8), though the more alkaline are generally thought to be better. They will also grow in anything except pure sand or pure clay, though well-aerated, fertile soils of good depth are to be preferred. The soil analysis will tell you about the levels of key nutrients and trace elements such as nitrogen, phosphorus, potassium, iron and boron. This, and your consultant, will give you an idea of what you need to add to make sure that the trees are growing in optimum conditions. For example, it may be a good idea to add lime to the soil before planting, to increase the availability of calcium and to reduce acidity, or to add gypsum to open up a clay soil. The other elements of plant nutrition we'll look at in Chapter 10.

The next step is to dig a hole so that you can see how the soil changes with depth. On a flat site where the underlying geology doesn't change significantly, one hole may be all that is required, but in more complex situations you may need to dig a few so that you can track changes. The soil profile will tell you how deep your soil is and what's going on underneath the olives, particularly in terms of drainage. The hole needs to be good and deep – preferably a couple of metres – and is best dug with a mechanical digger (especially if you've got a few to do!) If the hole fills with water, and the water table is less than 1 metre from the surface, then the site will require drainage, or won't be suitable. A hard, compacted layer below the surface, known as a pan, can also impede drainage (and root growth), and if one is present it may be necessary to break it up.

Site selection

Climatic and soil factors are crucial. Get those wrong, and the chances of your grove proving economic are greatly reduced. Local knowledge is also very helpful, or at least an understanding of the worst that nature can throw at you. We planted our first trees (oaks and hazels, not olives) just as the worst drought for at least 40 years got under way. The experience of that summer made us radically revise our plans for irrigation, and underlined the need for shelter. If we'd been 'lucky', and had a few 'normal' years before the drought struck, we might have been badly prepared and suffered significant losses. Now we know that we can survive the worst that El Niño can send, we feel a great deal more secure (though La Niña's a bit of a worry).

It's also comforting to see neighbours growing olives commercially, or at least trying to. Someone else thinks the climate's good, and likes the business, and there's often strength in numbers. The presence of other growers also indicates that shared expertise is likely to be available, including informed agricultural consultants, and in the longer term, there may be enough production to encourage the establishment of a local press. If you are pioneering a totally new area, or one with a known problem such as occasional severe late spring frosts, life may be a great deal less comfortable and more costly.

Aspect

So if the climate and soil is right, shelter is in place or can be planted, and waters available, what other factors need to be considered? The aspect of the site is clearly important, and not just to growers in cool climates. It should be north-facing, to make the most of the summer sun, and shouldn't have significant shading early or late in the season. A flat or mildly sloping site will make working easy, but in frost-prone areas a sloping site that has somewhere for the cold air to drain to (and where it won't collect) is obviously important. Slopes that are too steep for tractor work will demand a lot of manual labour unless they're terraced, which may be very expensive.

Access

Easy access is another key requirement. You'll need to be able to get the normal range of farm machinery onto the site, and later, if you plan to use mechanical harvesters, those bulky beasts will need to be able to get in among the trees. If this means cutting roads, or improving existing ones, this expense will also add to start-up costs.

Bureaucracy

One final important point. Your local authority may require you to apply for permission to change the use of your land. If you are deemed to be in a sensitive landscape, for ecological, cultural or aesthetic reasons, you may find that you have to meet certain conditions if your grove is to get the go-ahead. Planning authorities may also have restrictions about trees being planted close to roads, or require you to account for any run-off into local watercourses. And if you plan to sink a borehole to extract water, you will certainly have some bureaucracy to work through.

FINANCIAL PLANNING

Sentiment and romance can carry you a long way, but ultimately it will be money that decides whether your grove flourishes or becomes yet another exciting, innovative agricultural failure. You need to consider the various costs that go into establishing the grove, from buying the trees to the land itself, and the years of maintenance waiting for the first economic crop. If you decide to press your own fruit, you'll need to factor in the capital cost of a press, and all that goes with it. You'll require a business plan for the bank manager, and reserves for when things go wrong. And it should still be fun…

Before making any decisions on planting an olive grove, you should talk to your accountant. There may be important tax considerations, depreciation and all sorts of other things to consider before you spend any money or plant a tree. Your accountant should also be able to help you build a business plan, if you need one to raise money. A few dollars spent on professional financial and tax advice early on could save a lot of money in later years.

If you have the expertise, you'll find it very useful to prepare your own spreadsheet on a personal computer, summarising the costs discussed below, and projecting the yields and incomes into the future. A sophisticated Excel model is available with Michael Burr's *Australian Olives*, or from the New Zealand Olive Association, but putting together a simple outline of revenues versus costs is not too taxing if you know how to handle a mouse.

Olive grove projections

	Year	1	2	3	4	5	6	7	8	9	10
	Number of trees	350									
Costs	Planting	$3,500	$500	$50							
	Management	$2,500	$2,500	$2,000	$2,000	$2,000	$2,000	$2,000	$2,000	$2,000	$2,000
	Harvesting	$0	$0	$0	$0	$1,250	$1,250	$2,500	$2,500	$2,500	$2,500
	Annual cost	$6,000	$3,000	$2,050	$2,000	$3,250	$3,250	$4,500	$4,500	$4,500	$4,500
Revenues	Fruit yield per tree (kg)	0	0	0	0	5	15	25	35	35	35
	Total fruit (kg)	0	0	0	0	1750	5250	8750	12250	12250	12250
	Price per kg	$0.00	$0.00	$0.00	$0.00	$0.75	$0.75	$0.75	$0.75	$0.75	$0.75
	Net revenue ($/1)	$0	$0	$0	$0	$1,313	$3,938	$6,563	$9,188	$9,188	$9,188
	Revenue	$0	$0	$0	$0	$984	$2,953	$4,922	$6,891	$6,891	$6,891
	Profit/(Loss)	($6,000)	($3,000)	($2,050)	($2,000)	($2,266)	($297)	$422	$2,391	$2,391	$2,391
Totals	Total cost	$6,000	$3,000	$2,050	$2,000	$3,250	$3,250	$4,500	$4,500	$4,500	$4,500
	Total revenue	$0	$0	$0	$0	$984	$2,953	$4,922	$6,891	$6,891	$6,891
	Profit/(Loss)	($6,000)	($3,000)	($2,050)	($2,000)	($2,266)	($297)	$422	$2,391	$2,391	$2,391
	Cumulative Profit/(Loss)		($9,000)	($11,050)	($13,050)	($15,316)	($15,316)	($15,191)	($12,800)	($10,409)	($8,019)

COSTS

Upfront costs

The most obvious upfront cost is that of the land. If you plan to use land you already own, then you may choose to ignore this cost, but if you're paying a mortgage, you should certainly set the relevant proportion of the loan cost against the olive grove. If you're buying land specifically for an olive-growing venture, then this will certainly need to be factored in. Of course, land is always an asset, but it's very difficult to predict whether the asset value will increase, or whether a mature olive grove on the land will actually increase its value. It should, but by how much is anyone's guess.

You'll also need to allow for any professional advice that you use in selecting the land, including testing the soil and drainage, and any research that you undertake into the general suitability of the site.

The next major cost to consider is the provision of irrigation water to the site, and laying down the network of pipes to distribute it to the trees. If you are lucky, there will be a spring in the hill behind the grove, and all you will need is a tank and some pipes to gravity-feed water. The less fortunate will have to sink a well and pump water a long way. Getting an electricity supply to the pump may also prove expensive. In my case, despite siting the tank and pump *underneath* an existing power line, the power company still insisted on putting in a new pole at considerable expense (to me).

The equipment that you use in the grove also needs to be purchased and accounted for. See Chapter 10 for a discussion of what you might need.

Planting costs

You'll need to allow for the cost of the trees and their delivery to the site, buying good stakes, the irrigation drippers or sprays, the ripping and other site preparation, any fertiliser you apply, tree guards or pest protection, tree ties, and the labour cost of putting the trees in the ground. The party afterwards may, or may not, be an allowable business expense.

Annual costs

After the first year, with the major capital expenditure out of the way, you need to provide for the cost of maintaining the grove. This amount should cover the labour cost of pruning, any fertilising and soil and leaf sampling you carry out, plus the time you spend in keeping an eye on the trees and their development. Don't be tempted to underestimate the time you'll spend in the grove, adjusting tree ties, maintaining the irrigation network, crushing leaf-roller caterpillars.

You should also be depreciating the capital value of the equipment you've purchased. The rates you use for this, and what exactly can be depreciated, will depend on the tax rules that apply in your circumstances. Your accountant can advise.

Finally, when you begin to produce a crop of fruit, you will need to allow for the costs of harvest. This will depend on how you choose to pay your harvesters – by the hour or by the kilogram. In small groves or when trees are small (or very large), it may be better to pay by the hour.

Revenues

This is where applied guesswork comes into play. You'll need to make realistic assumptions about when your trees will begin to produce, and how much fruit they will yield. If you're over-optimistic, you will be preparing yourself (and your bank manager) for a nasty shock.

Your trees could begin to produce enough fruit to be worth picking in year four. How much, and how good it is, will depend greatly on your local conditions and how well the trees are growing. In a conservative projection, you might ignore year-four production completely. Some Australian projections suggest that you'll get 10 kilograms of fruit per tree in year four, rising by 10 kilograms per tree per annum to reach 70 kilograms per tree by year 10. Others suggest a yield of 35 kilograms per tree in year seven, with modest growth thereafter, while still others target 30 kilograms per tree, and regard anything extra as a bonus. The latter approach is certainly conservative, but it is far better to understate your income and overstate your costs and still show a profit, than to fall into a financial black hole when costs come in way over budget, and revenues fail to meet expectations.

One more imponderable: to what extent will alternate bearing be a problem in your grove? Young trees are markedly less prone to the problem of off-and-on years than older trees, and irrigation helps to reduce its effect, but it will almost certainly begin to have an effect on your production in the second or third decade of the grove's life. It might take only one late spring frost, or long hot summer, to push an entire region's trees into a synchronised pattern of alternate bearing. It's wise to be conservative in your projections.

The next piece of crystal ball gazing is to estimate the price your fruit is going to fetch, or what your oil will command in the marketplace. Neither is exactly straightforward, but it's certainly easier to assume that you'll sell your fruit, than to try and guess what presses will charge, what bottles are going to cost, and what the market will pay for your oil in five or six years' time.

At the time of writing, Australian presses and co-operatives were expecting to pay up to A$1 per kilo of fruit, while New Zealand prices ranged from NZ$2 per kilo for Barnea to NZ$1 for anything else. As the market becomes more sophisticated, however, and the volumes of fruit being sold increase, the price you achieve will depend on oil content and acidity. Prices will also inevitably fall as volume of production grows, so the cautious might take $0.50 as their expected selling price, and perhaps use $0.75 or $1 as the basis for 'high' income projections.

If you intend to produce your own oil, the procedure is more complex. You have to take into account the pressing cost, and the costs of bottling and marketing, as well as the expected revenues you achieve. It is very tempting to extrapolate optimistically: 200 trees at 35 kilograms per tree gives 7000 kilograms, at 20 percent oil yields 1400 litres of oil, which I can sell at $50 per litre, giving me a revenue of $70,000. Not bad from about 0.8 hectares of what was prime sheep land.

In your dreams.The above calculation ignores the costs of pressing and bottling and marketing the oil, and doesn't allow for the retailer's margin. If you plan your own press you can add in a raft of extra costs (see later). Since all these factors are very difficult to determine at this stage, the best advice I can give is to talk to any local presses about their charges, to winemakers about their bottling costs and then assume that you'll be giving retailers at least 25 percent of what the customer pays for your oil. Then look in the shops to see what is available, at what price, in your proposed market. If you can work back from your competitor's retail price and still make a profit, you're in the right ballpark.

For example, at the time of writing, New Zealand supermarkets were selling several different brands of single-estate type European olive oil at between $13 and $15 for a 250-millilitre bottle, or approximately $50 to $60 per litre. These are reasonably good oils, but they won't be as fresh as locally produced oil, and are certainly not top labels. To compete in the same market, you will have to be able to do everything, from growing the fruit to promoting the oil, and end up at the same sort of price. You should also bear in mind that, by the time your grove is producing fruit, many more oil producers in Australia and New Zealand will be marketing their oils, and the shelves could be getting crowded – another factor likely to depress returns. Your oil will need to be good enough to compete in what, if some current oils are anything to go by, is going to be a very high quality marketplace.

Despite all those caveats, there's no doubt that selling your own oil will earn you a better return on your investment than simply selling the fruit. In the jargon, you're adding value, and therefore will earn more. You're also taking a higher risk – something that should also justify a higher level of return.

Viability and returns

Your grove should begin to make an operating profit fairly early in its life, but that doesn't mean that it's financially viable. If the level of capital required to get it going is high, then the returns will also have to be high. At the very least, your earnings should exceed the money you would have earned if you had the same investment sitting in an interest-earning bank account. But that ignores the intangible benefits of owning and running a grove, being in an exciting business and having an excuse for eating lots of gourmet food. The value you place on those aspects may allow you to run your grove in a way that pure businesspeople (or their accountants) might find very sloppy or unattractive. Only you can decide how to spend your money, and what constitutes a reasonable return. And if all you want to do is to earn more money per hectare than growing wool, you should be on to a winner.

Setting up a press

As we saw in Chapter 5, there are many good reasons why an olive grower might want to have his or her own press. This, however, adds an extra level of risk and requires a substantial extra investment. The smallest feasible setup, using an Italian-made Oliomio machine, will cost in the region of A$20,000/NZ$23,000. That could probably be housed in an existing farm building, but by the time you get into larger hydraulic presses you'll be moving nearer to the $100,000 mark and will probably require a new building to house the whole setup. Large centrifuge systems will need a small factory building, and hundreds of thousands of dollars of investment.

The economics therefore depend on making a carefully calculated decision about your commitment to the business. A larger grower aiming at the boutique single-estate market might be able to justify the expense of establishing a press, as Mike Ponder has done. Others will prefer to let someone else spend the money and take the risk, at least in the meantime.

SELECTING YOUR TREES

Before you select the cultivars you're going to plant, you have to decide what you're going to do with your olives when they're in full production. There are two obvious courses: to grow olives for oil, or for pickling as table olives. That simple choice will determine much about your choice of cultivars. But that's only the first step. You then need to think about the market for your olives. Are you going to sell them to a local oil producer or a table olive processor? If so, you'll have to talk to them about the varieties they'll want to buy.

If your nearest producer specialises in oil made from the Barnea variety, you may be able to expect a premium for Barnea fruit. On the other hand, a producer from the 'Tuscany's best' camp will be looking for Frantoio or Leccino fruit. Similarly, a table olive producer may prefer Kalamata or Spanish Queen to other varieties. A little research in the very early stages of planning may save some expensive mistakes and make all the difference between selling your crop at a decent price or having to accept whatever's offered. It's also possible that a local oil or table olive producer may be prepared to offer a contract to buy your fruit (common practice in the wine business), though at the time of writing this is more likely in the burgeoning Australian market than in New Zealand. Australia is also seeing the establishment of producer co-operatives and grower-owned pressing companies, so it will be well worth talking to other growers in your area to find out what's going on.

But life may not be that simple. In many areas there may be no local presses or processors. If there is no critical mass of fruit, or if production is not expected to begin for a few years, then it may be difficult to get sensible guidance about what to plant. You may have to take a bit more of a gamble, and try to guess what will be required in a few years' time.

To simplify matters, you may decide that you want to establish your own small (or large) press to process your fruit, and then bottle and market your own oil. This increases the capital outlay considerably, and will demand a great deal of attention to sales and marketing, but does mean that you will be in control of your own destiny. This was certainly the vision that my wife and I had: try to make the finest olive we could, and to try to echo the fine oils of Europe, and in particular Tuscany. That has turned out to be a trendy thing to do (though we didn't know it at the time). We didn't want to get involved in producing large quantities of fruit, but we did want to grow fruit of the highest quality and press oil that would stand comparison with the best. With luck, we may be able to bottle our oil and sell it in sufficient quantities, and at a high enough price, to make the land and trees earn their keep.

The range of cultivars available for you to choose from will vary, depending on where you live and which nursery you talk to. In general, Australia has more varieties available (and in greater quantities) than New Zealand, reflecting the fact that Australia has had an olive-growing industry since early colonial days and that the country is in the middle of a real olive boom. New Zealand's business is smaller, but on both sides of the Tasman you should be able to find a nursery that can supply you with most of the major, internationally planted cultivars.

Fashion

I hinted earlier that fashion has its part to play in the business, and the current fashion is for Tuscan cultivars. This isn't surprising when you consider that the fine green oils of the Tuscan hills are the aristocrats of the trade. Unfortunately, in New Zealand and much of Australia, Tuscan varieties have little or no track record of commercial production. Plenty of experienced people will advise that the best thing any new grower can do is to plant varieties that are proven producers in your local conditions. New Zealand's first commercial oil producer, Mike Ponder, is quite adamant that only Barnea and Manzanilla are proven producers in New Zealand. Other experienced growers might add one or two other varieties, but nobody can point to a grove of mature Tuscan trees producing commercial quantities of fruit. Nor can anyone yet say what flavours the oil will contain. In 1998, Ponder produced limited quantities of a single pressing of Frantoio (available only from the estate shop), which has at least some of the character it is supposed to contain. As the years go by, we may discover that the flavours expressed in New Zealand-grown Frantoio may be very different from its Tuscan cousin, and different again from the Australian version of the same thing.

And even if Australia and New Zealand are capable of producing oils similar in character to a Laudemio or Coltibuono, that Tuscan bite and pepper may not be to everybody's taste. It may be that the Tuscan flavour is no advantage, or even a liability.

If you can't follow the advice of someone who wants to buy your fruit, and wish to establish your own oil 'label', you will need to think this question through and arrive at your own best guess. The question of scale is also important. If, like me, you're aiming at the boutique producer category, then you'll need to have some idea of how your oil might fit into the market. Will it be possible to have a diversity of small and medium producers, as in the wine business, each with their own unique blends and claims to fame? I certainly hope so. If you're aiming at a larger market, perhaps a place on the supermarket shelves, then a blander, mellower oil may be just what is required.

Pollinators

Some cultivars aren't self-fertile: they need to be pollinated by another cultivar. Barnea, for example, needs Picholine (the officially recommended pollinator) or Manzanilla (the one that just happened to work in New Zealand). A 'Tuscan' grove will need Pendolino, to get the best yields from Leccino and, to a lesser extent, Frantoio. Most authorities recommend that pollinators should account for about 10 percent of your grove. In general, to get a good range of pollens blowing around, it is best to plant a good range of cultivars – at least four or five – and to make sure that the pollinators are spread out through the grove.

We've already looked at the huge diversity of olive cultivars in use around the world (Chapter 3), so this section is going to concentrate on those that the Australian and New Zealand grower is likely to encounter in the nursery. Before making any selections, you should be thoroughly familiar with your climate, the varieties being grown by neighbours and the varieties in demand from local or regional processors. My comments are based on information published by various nurseries in the two

A ten-year-old Frantoio tree at the Marlborough Olives nursery, near Blenheim.

countries. They often differ in their claims about yield and vigour (perhaps reflecting different genetic stock), so I've tried to steer a middle course. I've organised the cultivars by national country of origin and into oil and pickling varieties. Quite a few cultivars are considered to be dual-purpose, and not just because they're not much good for either. Kalamata, the famous table olive sold by name at delis around the world, is also pressed for oil, and it's an oil that some experts rate very highly (On the other hand, some 'Kalamata' oil is oil pressed from Koroneiki olives in and around the town of Kalamata, and who knows what the experts actually tasted.)

OLIVES FOR OIL

Growing olives for oil is all about getting the maximum oil yield per kilogram of fruit, at the lowest possible acidity. That will determine the price you receive for your crop, or the amount of oil you have to sell. Flavour will come some way down the list, and is, to a great extent, an unknown for any Northern Hemisphere variety not currently producing down here. Equally, the percentage oil figure quoted should always be taken with a pinch of salt. Not only will it vary from year to year, but also from site to site, and will also depend on the way the trees are managed.

ITALY

Frantoio is one of the workhorses of Tuscan oil production, renowned for the flavour and quality of the oil it produces. The tree bears good crops of smallish olives that ripen late in the season, and yields oil at somewhere between 15 and 18 percent. It's a fairly vigorous tree, and cold tolerant. Although not yet widely planted in Australia and New Zealand, it is very fashionable. The Australian *Corregiola* is said to be the same, or from the same family.

Leccino is the other major Tuscan variety, producing good low-acidity oil that is said to be sweeter than Frantoio. Fruit ripens earlier than Frantoio. One of the most frost hardy cultivars, but it needs a pollinator – Pendolino is the usual choice. In New Zealand it is said to be more vigorous than Frantoio.

Pendolino (Pendulina), as its name, 'little pendulum', suggests, does have a characteristic pendulous form. Branches tend to hang down, and they're easily spotted in an otherwise fairly upright grove. A pretty good oil cultivar in its own right, said to yield up to 22 percent. Usually planted as a pollinator for Leccino.

Moraiolo is well known, and planted throughout Tuscany. This is another frost hardy cultivar that produces good oil. It is said that the fruit is very firmly attached to the tree, and so has to be hand picked. The fruit can also be left on the tree for some time without significant deterioration. Still early days to judge its performance in New Zealand, and it is not yet introduced to Australia.

Carolea is widely planted in Italy, and is said to be a good oil producer. Still early days to judge its performance in New Zealand, and it is not yet introduced to Australia.

FS-17 and DA-12 are only available in Australia at the moment. FS-17 is a unique dwarf cultivar and DA-12 a dwarfing rootstock developed in

A healthy Picual tree, doing well a long way from its origin in Spain.

Italy by Professor Giuseppe Fontanazza. The basic idea is to grow small trees that can be harvested and pruned by special machines that straddle the trees, much like other kinds of fruit pickers and pruners. If the trees are planted close together, yields are said to be as good as in conventional orchards, with a very low labour cost. This is all cutting-edge stuff, and some people remain resolutely sceptical about the long-term prospects for this style of olive growing. Nevertheless, for intensive cultivation on a large scale, the cost savings could give an enormous competitive advantage. Definitely one to watch.

SPAIN

Picual is the most important Spanish oil cultivar, with over 600,000 hectares planted. It also accounts for more than half of all new plantings. It is a vigorous grower, producing early crops with oil yields claimed to be as high as 23 to 27 percent, tolerant of wet soils and reasonably cold hardy. The cultivar known as Nevadillo Blanco in Australia is almost certainly the same, or very closely related, and is said to need good winter chilling.

GREECE

Koroneiki is a tough tree that is the primary oil cultivar in Greece, producing very good oils. It is said to have a low chilling requirement,

Koroneiki flowers about to open: mid-November in Canterbury.

making it suitable for warm areas, and is also tolerant of hot, dry summers. It grows quickly and is an early cropper. Fruit is small with a high percentage of oil. Tolerant of wind and so can be used for shelter, and very resistant to peacock spot.

FRANCE

Picholine is also a good pickling olive with a flesh-to-pit ratio of 5 to 1, producing crops of medium-sized fruit that have a good flavour. The official pollinator for Barnea, it is said to be adaptable to a wide range of soils and cold tolerant.

Verdale is considered a dual-purpose variety in its southern French homeland, producing medium-sized fruit with a reasonable oil content but a large stone. The trees tend to remain small, and are prone to alternate bearing. This cultivar is common in older plantings in Australia.

NORTH AFRICA AND THE MIDDLE EAST

Sourani is a vigorous cultivar, of Syrian extraction, reputed to have a high oil yield. Not an early cropper, at least not in New Zealand conditions. Might do better in warmer areas.

Chemlali is from Algeria, but also planted in Greece. Has a lot in common with Koroneiki, but has larger fruit that ripen earlier, and doesn't come into production so quickly. Said to be resistant to peacock spot.

ISRAEL

Barnea was developed in Israel by Shimon Lavee and his team for intensive irrigated groves using mechanical harvesting. It grows vigorously and tall, and will come into production quickly if provided with a pollinator. Picholine is the official recommendation, but Manzanilla has been found to work just as well in New Zealand. It is said to yield heavy crops at about 20 percent oil, and can also be used for pickling. It has been extensively planted in New Zealand, but has only recently become available in Australia.

Souri is said to be a slow-growing tree that produces medium-sized fruit with a very high oil content. It is sturdy, but susceptible to peacock spot

Nabali Mouhasan is a fast-growing tree that can crop very early in life. The fruit is medium-sized with a good percentage of oil (up to 18 percent), but can also be pickled. Said to be susceptible to peacock spot.

UNITED STATES

Mission (Californian) arrived in California with Spanish missionaries, and is definitely of Spanish extraction – possibly related to Picual. Produces good oil (up to 18 percent) from medium-sized fruit, but is mainly used for pickling. It's very prone to alternate bearing, but has very good resistance to cold. Is not now regarded as a major commercial cultivar in its homeland.

The Israeli cultivar, Barnea, tends to form a tall tree, making it suitable for machine harvesting.

AUSTRALIA

Because so many different varieties arrived in Australia with the early settlers, and under a variety of names, there are potentially a huge number of cultivars available. Gene testing is under way to identify as many different cultivars as possible (by comparing them with 'known' trees in the IOOC's germplasm collection). The best local selections will also be offered for sale.

Mission (WA) is not the same as the Mission from California. This Mission comes from the trees used by the monks at the New Norcia monastery in Western Australia to make their award-winning oil.

Paragon (Mediterranean) produces heavy crops of smallish fruit with a high oil content (23.5 percent), and has been widely planted in Australia. Good in warmer areas, and a vigorous grower. Apparently popular as a garden tree for home pickling enthusiasts. Recent DNA tests suggest that this tree is identical to Frantoio, and may in future be sold as Frantoio, or Paragon-Frantoio.

Hardy's Mammoth produces healthy crops of large fruit very suitable for pickling, but also with an oil content of up to 23 percent. Said to be a tough, vigorous tree.

Tiny Oil Kalamata is probably only a distant cousin of the real thing, producing similar-shaped fruit, though as the name suggests, much smaller. The fruit can appear like grapes on the branches. Oil content is good.

NEW ZEALAND

A number of keen individuals are trialling selections from old local trees which seem to be doing well, and some are finding their way into nurseries. These include J1 and J2, GB01, One Tree Hill, Rakino and Super. There's even one called Pooch. The hope is that, among these trees, a few will prove ideal for New Zealand conditions. In the meantime, most people are concentrating on planting more widely recognised cultivars.

OLIVES FOR PICKLING

A completely different set of attributes is looked for in olives destined for the processor's vats. The fruit should be a reasonable size, with a good flesh-to-pit ratio (that is, a small stone in a large olive), resistant to bruising and should retain its texture and flavour through the pickling process. In this market your choices are likely to be limited by what your local processors want, unless you plan to go into the processing business yourself, in which case you may be able to experiment a little more. The New Zealand table olive grower needs to consider this last point very carefully, because at the time of writing there is no olive processor of significant size in the country.

ITALY

Uovo de Piccione ('pigeon's egg') bears large fruit, hence its charming name. Opinions differ as to whether it's vigorous or slow-growing (my two aren't growing quickly) but, like Kalamata, it can be difficult to propagate, and is usually rather expensive.

FRANCE

Tanche de Nyons is the only olive to have its own *appellation controllée* label, and is widely thought to be France's finest pickling olive. Not currently available in New Zealand, but should soon be available in Australia.

SPAIN

Sevillano (Spanish Queen) is a slow-growing tree producing very large fruit (up to 13.5 grams), with a reasonable flesh-to-pit ratio but relatively low oil content. Rarely planted commercially these days because of a number of problems, including a susceptibility to shotberries, 'soft nose' (a nutrition-related distortion of the end of the fruit) and because the fruit bruises easily. It has now been largely replaced by Manzanilla.

Manzanilla is the world's most popular pickling olive, widely planted in Spain and California. It also produces a very good quality oil, but at yields of under 15 percent. A relatively small tree, it crops heavily and comes into production early, but is more prone to alternate bearing than some. A proven producer in New Zealand, and popular in Australia, where it did well in a 14-year trial at Mildura in the 1960s and 1970s. It is sometimes said to need Uovo de Piccione as a pollinator, but few growers seem to have problems with fruit-set.

Hojiblanca is a hardy and cold tolerant cultivar, widely planted in Spain for both oil and pickling. Tolerant of high pH soils, and will stand drought. Fruit is large with a flesh-to-pit ratio of between 5 and 6.5 to 1, but the oil content is relatively low.

GREECE

Kalamata olives are sold by virtually every delicatessen worth the name, and they deserve their fame. Harvested when fully ripe, they pickle

Ripe Kalamata fruit at the Coriole winery in McLaren Vale near Adelaide.

beautifully and have a very good flavour. This olive can also be used for oil, with yields said to be between 17 and 19% percent (others put it higher). It's a medium-sized tree with medium-sized crops, but is difficult to propagate and therefore expensive to buy.

Jumbo or *King Kalamata* is probably not a true Kalamata, but that's what they call it in Australia. The fruit is the same sort of colour and shape as Kalamata, but much larger at around 10 grams per olive. Another tree that is difficult to propagate, and so more expensive in the nursery. Not yet available in New Zealand.

Volos is said to produce large fruit suitable for green or black pickling, but also has a relatively high oil content. Trees are being trialled by Mike Ponder in New Zealand.

NORTH AFRICA

Barouni is a North African cultivar said to be better for pickling than for oil, because of a relatively low oil yield. The fruit can be large, with a reasonable flesh-to-pit ratio. Said to be cold tolerant, and grows into a spreading form, making it easy to harvest.

Nab Tamri is a large tree that produces moderate to heavy crops of 10- to 11-gram fruit with a good flesh-to-pit ratio.

ISRAEL

Kadesh is an unusual olive specifically developed to have a low oil content, designed to appeal people trying to follow a low-fat diet. The fruit is medium-sized, with oil content as low as 3 percent, and is said to taste good.

UNITED STATES

UC13A6 (California Queen). UC stands for the University of California, which developed this cultivar. It's a medium-sized tree, giving good crops of large round fruit – up to 17 grams has been recorded in Australia, though 11.5 grams is more normal. Has a good flesh-to-pit ratio, and is generally picked green.

AUSTRALIA

SA Verdale (the SA stands for South Australian) is a selection from 'old' Verdale, and produces large fruit suitable for pickling. The tree is said to be sturdy and to stand up to wind well.

Choosing a nursery

Olive nurseries are wonderful places full of little trees, all seemingly identical, growing rapidly towards a saleable size. And the people running them are also often wonderful – enthusiastic, knowledgeable, helpful and paragons of horticultural virtue. You hope. Choosing a nursery to supply your trees is a critical part of the process of getting your grove in the ground. After all, you're only (we hope) going to have one shot at planting that paddock, so it makes sense to ensure that you get the best possible trees, the best possible advice, and a good price.

The first thing to consider is the source of the genetic material used in propagating the trees. A good nursery will be happy to show you the mother trees, or to tell you where they got the material they used. This is particularly important when you consider that a cultivar may have several different names, or several cultivars the same name, or may be a local selection from an old tree whose name is lost in the mists of time. The mother trees should be healthy, and free from obvious pests and diseases. The bigger

One day these will make a lot of oil. A small part of the giant Olives Australia nursery in Queensland.

they are, the longer the nursery will have been in business, and although that in itself is not a guarantee that they're any good, it should mean they've been around long enough to learn from their mistakes.

Listen carefully to what nursery people have to say about planting and looking after your trees. If they know their stuff, they will be familiar with your climate and the sorts of cultivar that might be expected to do well. With luck, their comments shouldn't be too far away from those in this book. If they differ, don't be afraid to ask why. Learning about olive trees is a continuous process, especially in the Southern Hemisphere, and some things are matters of personal preference.

Look at the trees for sale. They'll probably come in several grades, from large, healthy trees at large healthy prices, to runty little things costing less. Smaller trees may simply not have had time to grow – perhaps because demand has not been high – or they could be slower-growing varieties, so don't necessarily reject them out of hand.

The price you pay for your trees is obviously important – especially if you're buying thousands of them – but don't be tempted to put getting a bargain ahead of getting top quality. It might be possible to save a few dollars per tree, but if that means that the trees have a higher failure rate, or take longer to come into full production, or are a lesser variety with a lower oil content, then you will pay in the long term.

One final warning: don't expect to be able to make one phone call and take delivery of your trees a few weeks later – especially if your order is large. At the rate the business is currently expanding, many nurseries will be back-ordered.

The last resort

Disaster has struck. You've planted the wrong cultivars, and can't sell the fruit, or the oil isn't living up to expectations. Do you have to pull the whole grove up and start all over again? With luck, and some skill, the answer could be no.

Olives are good at accepting grafts, so if you have healthy trees with well-established root systems, it may make sense to prune away the whole crown of the tree, and graft another variety onto the remaining trunk. This sort of thing – called top working – has been done for thousands of years in Europe. Many of the gnarly, hollowed-out trees of the Mediterranean may well have been grafted repeatedly as new varieties arrived on the scene or as fashion changed. It's a relatively simple, if labour-intensive process, but if it goes well you should be back in production before any new young trees have grown enough to produce a commercial crop. And it saves the heartache of ripping up all those trees you lovingly planted...

Newly top-worked tree at Burr's Beetaloo Olive Grove. The grafts are inside the bag at the top of the stump, and the trunk has been painted to prevent sunburn.

PLANTING THE GROVE

Start a year before you intend to plant your trees. This is the counsel of perfection; I don't expect to meet the person who follows this prescription to the letter. Although I did many things well in advance, the weather and some irrigation problems conspired to make the final effort distinctly rushed. But it was done, and done reasonably well.

Layout

Getting the layout of your grove right is another of those things you can do only once. Once the trees are growing, it's going to be very painful to find that you've planted them too close together, or that one of the rows you can see from the house has a distinct kink in it. Choosing how to space your trees can be a rather controversial subject. There are rival schools of thought, and several factors to consider.

The first thing you have to ensure is that machinery can get into the grove, and has enough room to turn round at the end of each row (the headlands). If you're planning to hand harvest a smallish grove, then you may be able to get away with a relatively tight spacing, and modest turning room at the headlands, but a large grove to be harvested by shaking machines will need wide rows and lots of turning room. In Australia, the standard recommendation for mechanical harvesting is 6 to 8 metres between rows, and 5 to 6 metres between trees in rows, with at least 8 metres turning room.

Irrigated groves can be planted more densely than those that rely on natural rainfall, or get only occasional supplementation. Trees that are going to have to fend for themselves will need more room for their root systems, so they can get as much as water as possible when it's available.

Soil conditions are all-important for organic grower Peter Maroudas in his grove north of Adelaide. His trees are over twenty years old.

My trees are planted in rows 6 metres apart, with 5 metres between trees in the rows. This 6 x 5 layout is about as standard as they come in New Zealand, though I might have opted for 6 x 6 or even 7 x 7 if I'd had more land to play with. This would have given the maturing trees a bit more room to breathe and grow, and bigger trees mean more fruit (all other things being equal). This sort of planting density (6 x 5 is about 350 trees per hectare) is well within what European studies have suggested is the maximum for fully irrigated groves.

One of South Australia's olive gurus, Peter Maroudas, reckons that you should give your trees a 12 x 10-metre layout (or 12 x 8 if space is tight), and that's with some supplementary irrigation. His approach as an organic grower is to encourage the trees to get big, and he regularly harvests 120 kilograms off his best trees in a good year. He reckons that one tree producing 120 kilograms is better than two trees in the same space producing 50 kilograms each. This is anecdotal stuff rather than scientifically collected data, but if I had a lot of land to play with I'd be very tempted to follow his example – or at least to trial an area.

There's one other aspect to consider: whether you plant in squares or triangles. In theory, the most efficient use of space in a grid of trees is to pack them hexagonally. Each tree is surrounded by six others, and has the maximum amount of space for its roots to grow before meeting its neighbours. This is easily achieved. Plant every other row so that the trees are spaced halfway between trees in the neighbouring rows (see diagram). The alternative, planting in squares, is slightly less efficient, but may look neater. Much will depend on the shape of the land you intend to use, and how you choose to align the rows.

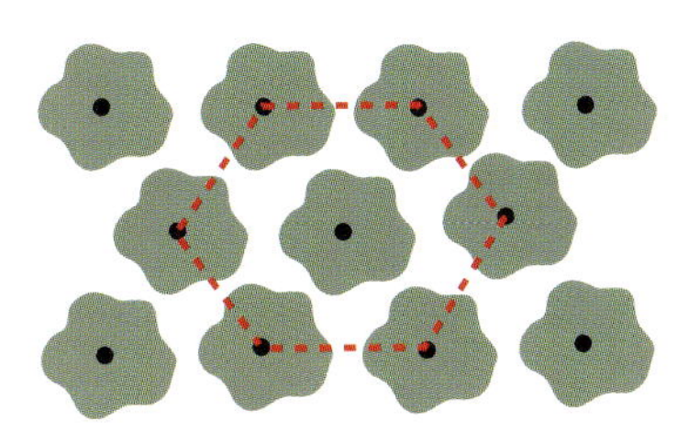

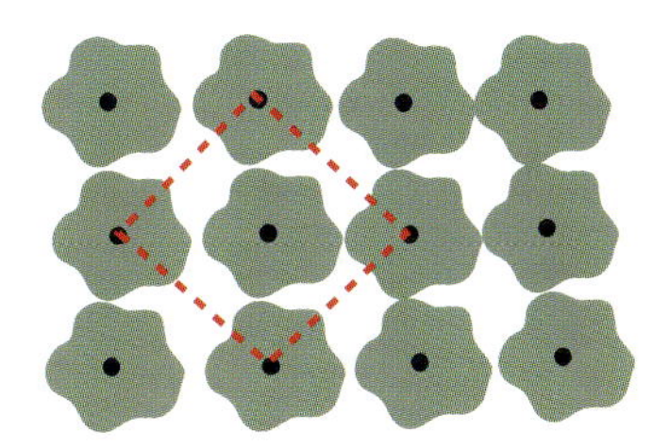

Staggering planting rows so that each tree is surrounded by six others (left) is the most efficient use of space, but some growers prefer planting in squares (right).

Most books advise planting crops in rows aligned north-south, to allow the maximum amount of sunlight into the trees. In a grove where the distance between rows is almost the same as the distance between trees, this can't be of any significance – you've got 'rows' going east-west as well as north-south. It might make sense for crops that are close planted in narrow rows, like grapes, but can't make much difference to olives. In some situations, you may want to plant your trees along the contours of the land, to help to control erosion, or as the first step towards a gentle sort of terracing. Another consideration might be to plant in rows facing into the prevailing wind, on the basis that, as the trees get bigger, the ones at the ends of the rows will begin to shelter the others.

Once you've decided on a spacing and layout, draw it out on a scale map. If your paddock is rectangular, this will be pretty straightforward. In my case, with a roughly triangular paddock, and the longest side of the paddock distinctly wavy, I used an aerial photograph blown up to 1:400

scale by a local colour photocopying company to establish the boundaries of the paddock. I then plotted my grid and worked out which trees were going where. The next stage was to translate my carefully worked out plan onto the paddock. This was done using wooden pegs as markers, with one end sprayed with fluorescent orange paint so that they would be easily seen when mowing, spraying and ripping. The rows were aligned roughly north-south using the shadow cast by a stick at noon (allowing for daylight saving time, of course). You could use a map and compass if you wanted to be really precise, remembering to allow for the deviation from magnetic north. This alignment meant that my rows would slant away from the edge of the paddock and gave me a slightly distorted version of hexagonal close packing. It also means that one of the alignments to be seen in the grove runs more or less northwest–southeast, the direction of the most damaging wind.

That all sounds terribly efficient; the truth was slightly different. When I marked the start of each row, I discovered that my map wasn't entirely accurate. I was getting more rows in than I expected. Then, when I marked out the length of the rows, it began to look as though I was getting more trees in some rows, and fewer in others, thus upsetting my careful calculations. Back to the drawing board. By the time I arrived at a fully marked-out paddock, and a map that agreed with reality, I must have walked several marathons and wasted half a book of graph paper. Those growers with fields laid out in a standard rectangular grid have life too easy.

Once you have a map, you can plot the positions that individual cultivars will have within the grove. The most important consideration, if you're planting cultivars that need the help of pollinators, is that you should space the pollinator trees through the grove so that their pollen can reach all the other trees. If you have a prevailing wind, you'll need to allow for that. Olive pollen will carry at least 20 metres, probably more, so you can either dot your pollinators through the grove, or plant whole rows of pollinators, perhaps every fourth row. The latter approach has the merit of making it easy to organise the harvest.

If you're planting varieties susceptible to wind, such as Leccino, then they should obviously be planted in the most sheltered part of the grove. Similarly, any notably wind-resistant cultivars could be placed in the windiest areas.

Mark all your tree positions on a map, and keep it in a safe place. Few olive cultivars are very distinctive; you may need to refer to your map at future harvest times, or when planning changes, or so that your children's children's children know what you got up to all those years ago. It might be a good idea to laminate the original and a photocopy, so that you can have one pinned up in the office for ready reference. I've also started to keep a computer record of my trees, so that I can log performance by variety and season. Setting up a database may be wearisome, and maintaining it a chore, but it may be very useful in the future, especially if you want to trial various cultivars or planting approaches.

Preparing the site

You'll know from your soil tests, and the soil profile, whether any specific preparation (such as adding lime, or improving drainage) is necessary. Assuming that's all done, the first thing to do is to rip along the rows. Ripping is a fairly brutal process: you need a hefty tractor and a very heavy sort of plough that dives deep into the ground, breaking up the soil and aerating it, and improving drainage. There are various kinds – triple rippers and vibrating rippers – but they all do the same basic job, which is to break up hard-packed soil so that young trees' roots can grow out easily. In heavy soils, it may be best to cross-rip – once down the rows, then across them at each tree position. Ideally, ripping should be done as long before planting as possible, to give the soil time to settle down. An autumn rip for a spring planting would be about right.

Weed control comes next. All the evidence shows that young trees grow much more rapidly if grass and weeds in their root zones are kept to a minimum. Growers with a leaning to the organic may find this fairly labour intensive, but most others will be happy to spray the grass and weeds with a glyphosate-based broad-spectrum weedkiller. These act only on the vegetation they touch, and are harmless once they touch the soil. A month or two before planting, spray a 1 to 2-metre strip down the rows. A boom sprayer mounted on the back of a four-wheel farm bike is perfect for the job. About two weeks later, everything should have died. The next step is to run down the rows with a rotary hoe, digging down as deeply as possible, to get a nice fine tilth. If you can wait for the weeds to germinate, and then spray again, so much the better – it'll save labour later.

Irrigation

If you're planning on installing irrigation, then now is the time do it. Unless you have considerable experience in plumbing plastic pipe and making

Jack the Ripper — a sort of winged plough that dives into the soil and 'rips' it, making it loose and easy for the new tree's roots to grow in.

pumps work, the main installation is probably best left to experts. The key decisions concern the amount of water you need to apply, and the method you choose to apply it. In the early years, the trees may need only small amounts of water, but as they get bigger they will obviously get thirstier. The water-holding properties of the soil also have a big part to play – a free-draining sandy soil will need more water than a clay-based soil to deliver the same amount to the tree, because the water will move out of the root zone much more quickly. Equally, be careful not to overwater, given the olive's dislike of wet feet. Too much water can also increase the risk of phytophthora, a fungal root rot that can kill the tree.

As a rule of thumb, newly-planted trees could require between 5 and 20 litres per week, depending on the soil, the ETP at the time and the method you choose to apply the water. Drippers are much more efficient than sprinklers, minimising surface evaporation and weed growth, but sprinklers may distribute the water over a wider area and encourage root growth in the top layers of the soil. One or two drippers per tree, placed 20 centimetres away from the trunk (*not* on top of the rootball) are all that's required. One sprinkler per tree will be enough. Later in the life of the grove, you should move the drippers further away from the tree, up to 1 metre on either side, or perhaps double the number of sprinklers.

Once your trees are planted, your most frequent activity in the grove for the first few years will be mowing the grass. Experience has shown me that burying irrigation lines with a mole plough makes that a lot easier. You can then mow across the rows as well as down them, without having to move the irrigation line or running the risk of chopping it into little bits. Drippers or sprinklers should be easy to avoid (but keep spares!)

If you aren't going to be able to monitor your trees on a regular basis, it may be worth installing a computer-controlled system, with a tensiometer (a device that measures the amount of water in the soil) to ensure that under- or overwatering isn't a problem.

Staking

With the irrigation lines laid (and marked on the map), the next step is to mark the tree positions. The simplest way to do this is to cut a rod to the distance between the trees in the rows, and then to walk down the rows pushing in a stake at each position. A visual check every now and again will reassure you that everything's lining up as it should (or not).

The choice of stakes is important. For my first lot of trees, I used untreated tomato stakes, wimpy little things that started to rot and crack and blow over after a year. For my olives, I used 50 x 50-millimetre treated pine stakes a nominal 1.2 metres long ('You're planting olives, aren't you?' said the bloke in the timber yard.) These sturdy beasts will last a good long time, and are rapidly becoming the New Zealand standard. In Australia, I've seen trees supported only by bamboo stakes, and these may be sufficient in sheltered locations. Anywhere where the wind can get strong enough to blow mature gums over, however, a good stake breeds confidence.

Once the stakes are marking the tree positions, you then have to go round and drive them into the soil. If dry soil makes this difficult, it may be an idea to install the drippers/sprinklers and give the positions a good soaking. You'll need to do this before planting in any case, and it certainly makes stake-driving a lot easier.

If, like mine, your grove has a complicated layout of cultivars, you might like to colour-code each stake according to the cultivar it's going to support. Builder's spray marking paint comes in a good selection of colours, and it's a simple matter to spray a blob of colour onto each stake. If you also colour code the pots or batches of trees as they arrive from the nursery, then you can be sure that each cultivar goes in the right place. The colour code will then be effective for a year or two.

Planting

With stakes in the ground and drippers/sprinklers in place, it's time to dig the holes. If you have a large number to do, you may wish to use a post-hole borer. If you do, look closely at the sides and bottoms of the holes. If the borer is producing a 'glazed' wall, it will be necessary to break this down with a crowbar (or similar), so that the roots aren't constrained. In small groves, with relatively soft soil, suitable holes can be dug manually without too much back-breaking work. In windy paddocks, make sure that the holes are downwind from the stakes, so that the trees are blown away from the stake, not into them.

Arrange to have the trees delivered a day or two before you plan to plant them. If the nursery arranges transport, it will also ensure that the trees are well packed. When they arrive, unpack them, and give them a good water. If you're collecting the trees yourself, be very careful that you don't expose them to wind damage. An open trailer at 80 kilometres per hour is just the same as a steady 80-kilometres-per-hour wind in the paddock. The trees will not enjoy it. If you're lucky, they'll just be stressed. At worst, they could be severely damaged.

Look at the size of the pots or planting bags the trees have arrived in. Make sure the holes are at least 10 centimetres deeper. If you want to add a little compost (make sure it's very well composted or rotted) or slow-release fertiliser, now is the time to do it. I used a pelleted product that contains the full range of nutrients, simply dropping one into the bottom of each hole. One tab should last the small trees through their first year in the ground.

The night before planting day, give the plants a good soaking. When you transport them to the paddock, make sure that they're not left lying around in the full sun all day without water. If the trees have arrived in black polythene planter bags, simply cut the bottom off the bag with a sharp knife, place the tree in the hole with the trunk about 10 centimetres away from the stake (to allow the trunk room to grow), and then gently

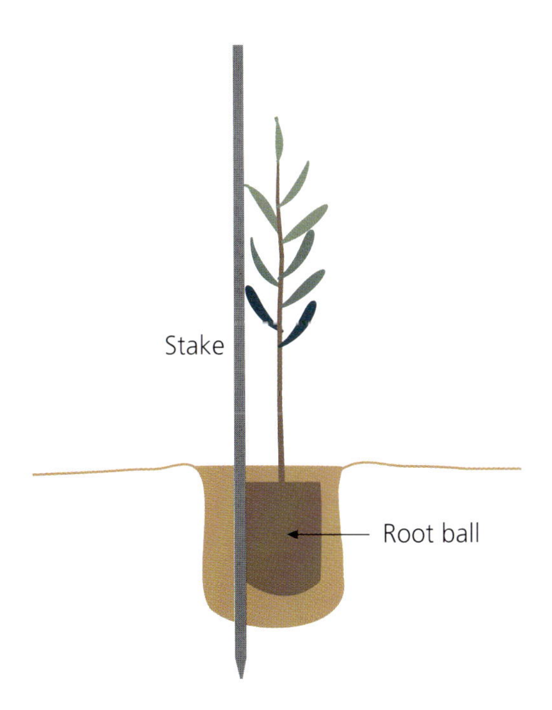

How it should be done: bury the rootball, and leave plenty of room for the trunk to thicken up.

Newly-planted Moraiolo, ready for its protective box. The drippers will have to be moved off the top of the rootball to prevent any rot problem.

pull the planter bag up off the rootball. If the trees are supplied in plastic pots, then simply place one hand over the top of the pot and tap the bottom to loosen the rootball. The pot should pull off easily.

Backfill the hole with soil, gently firming it in, and make sure that the soil covers the rootball by 5-10 centimetres. This gives the tree a good firm anchorage, as extra roots will grow out from the buried trunk. Some authorities recommend teasing the roots out of the rootball, suggesting that this encourages the roots to grow outwards. Others prefer to leave the rootball undisturbed, claiming that disturbing the roots damages them, and can set the tree back, as well as increase the risk of root rot in heavy soils. You can decide which camp you're in (I left mine undisturbed.)

If the trees arrive with nursery stakes still attached, these may be left on for the time being, provided that you check the tapes from time. You'll probably want to take them out within the first year. The tree's leader (the vigorous central vertical growing shoot) should be attached to the stake with some form of tree tie. There are lots of different kinds to choose from – anything from old tights to foam-covered wire. The main thing is that the tie should not cut into the bark, nor be so tight as to constrict the flow of sap. You need to monitor the trees closely during vigorous growth to make sure this doesn't become a problem.

If you're using any form of wind or pest protection, this should be slipped over the young tree and attached to the stake as soon as the tree is planted. Again, there are a wide variety of things to choose from. I used cut down KBC tree boxes, which provide excellent protection from wind and rabbits, and experience with my other trees suggests that they also provide some protection from frost. As soon as the tree has grown well out of the boxes they will be removed, and replaced with something else to protect against rabbit damage – probably a simple plastic sheath. If you have no wind or frost problem, a plastic sheath stapled round the base of the trunk may be all you need. Alternatives include a netting tube or, where pests are vicious things like sulphur-crested cockatoos or reinforced foil sleeves.

Once the planting is finished, switch on the irrigation and give the trees a good soaking. You are now officially allowed to breath a deep sigh of relief, gloat lovingly over the serried ranks of trees in your new olive grove and retire to the house for a long celebratory dinner.

The little Moraiolo disappears into its box. Within six months it was taller than the stake. We did something right!

Early management

For the first two to three years of the trees' lives in their new home, the most labour-intensive activity is weed and grass control. An area of at least 1.5 metres around each tree needs to be kept clear of weeds. This is easily achieved using glyphosate, though you must take care not to spray the young trees. Tree guards/boxes are a great help in this respect, though any weeds inside the box will enjoy the conditions just as much as the tree; they need to be pulled out by hand. A four-wheel farm bike with a motorised sprayer is a great help. Grass control is simply a matter of regular mowing, either with a tractor-mounted slasher, or a heavy-duty ride-on mower.

If you want to become an organic grower, you will either need to remove weeds mechanically, or use some form of mulch. Black plastic mats are certainly effective, but in moist conditions might increase soil humidity too much. Burning off weeds with a torch or flame wand is also effective, but would scare me rigid in the middle of a hot, dry summer.

There is no need to prune young trees. They should be left to their own devices until they have produced their first fruit, at least two to three years. The young tree arrives at a balance between the amount of leaves and the amount of root system. The leaves are the tree's factory, manufacturing all the things it needs to grow. Chop them off before the root system has had a chance to get really well established, and the tree will suffer a significant setback. If you're in any doubt, leave your trees alone.

The trees will need to be monitored regularly (a good excuse for an evening stroll, perhaps), to make sure that the tree ties are okay, that growth is proceeding nicely and that no pest or fungal problems are creeping in. In the second and subsequent years, it may be worth checking the trees' nutritional status by getting leaf samples analysed. This will tell you whether you need to add any specific nutrients, or just let the trees look after themselves.

GROVE MANAGEMENT

If your trees have been behaving themselves, and growing rapidly towards maturity, the workload in the grove is going to change from mowing and weeding to pruning, feeding and harvesting (the mowing and weeding never stop). Pruning the trees towards their desired final shape will have begun after the second growing season. As fruit amounts begin to increase, you'll need to check the trees' nutritional status and add fertiliser and trace elements as necessary.

Pruning

Pruning has two main objectives: to produce a tree of the desired shape and size, and to keep the main canopy of the tree fairly young and vigorous – and therefore productive. Along the way, you'll be eliminating any dead wood, increasing light penetration into the centre of the tree and perhaps also helping the tree to recover after damaging stress, such as cold or disease. How you get these things done, however, is as much art as science, and there's no substitute for experience.

Novice growers should look for practical help, perhaps by attending a workshop run by their local olive association, watching a video or just observing someone who knows what they are doing. It's very easy to be too timid – I used to be very cautious with my roses, but now they get a good hacking, and love me for it. On the other hand, being overzealous can reduce fruit yields; the olive bears fruit on last year's growth, so chopping too much off clearly won't do much good.

There is also a sort of Zen about pruning a tree. You need to look at it from different angles, consider its current shape and the shape you want it to be, look for branches that will co-operate with you, and (at the risk of

Unloved and unpruned for over 20 years, this dryland North Canterbury olive tree is still fruiting, but has been allowed to develop multiple trunks.

sounding a tad pretentious) you need to be at one with the tree. Like a picture, the shape you create with the tree's cooperation will be on display for a long time.

The natural shape of an olive is a sort of big shrubby bush with multiple trunks. Because it is basally dominant, that is, it likes to shoot from the base of a branch, if it is left completely untouched the fruit forms only on the outside of the tree, because light is required to stimulate fruiting. Although few growers like to leave their trees like this, there is evidence that this is just as productive as trees that have been extensively pruned into an 'ideal' shape. The fruit is also relatively easy to harvest by hand.

At the other extreme is an approach being studied by Italian scientists. They have been running an experimental grove for decades, using a sort of coppicing system. Every year, every 10th row is cut completely back to the base of the trunk at ground level. These trees will produce lots of shoots the following year, and will be fruiting again within four years. Because the trees never get very large, they can be planted fairly close together, and over a 10-year period the data suggest that yields are just as good as under any other pruning regime, while the labour costs of pruning are kept to an absolute minimum. You also get quite a lot of firewood every year!

Pruning is traditionally carried out in the months following harvest, when the tree is dormant. It is possible that a little light pruning may also be needed in early summer, perhaps to reduce fruit set and discourage a tendency to biennial bearing.

Two basic shapes — the bushy vase (left) and the modern monocone.

Anyone who has journeyed around the olive regions of the Mediterranean will have seen olive trees in all sorts of gnarled and twisted shapes. In the Southern Hemisphere, most of us are so new to the game that regional styles have yet to emerge (and may never). Generally, we're advised to follow a couple of distinct approaches: the bushy vase, or the monoconical shape – a bit like a Christmas tree. It depends on whether we plan to machine or manual harvest, or are using a specific cultivar. Barnea, for instance, tends to grow taller, more quickly than other cultivars and is closer to the monocone in natural shape than any other olive. This makes it ideal for machine harvesting – which is what it was selected for in the first place. It should be noted, however, that some experienced Barnea growers beg to differ. Other cultivars require more pruning to get into the monoconical shape.

The basic requirement for machine harvesting is that the tree should have between 0.9 and 1.2 metres of clear trunk between the ground and the lowest branches, so that the jaws of the tree shaker can get in to the trunk and get a good grip. This is fairly easy to achieve: just trim off the lowest branches gradually, starting after the second growing season, if the tree is really growing strongly, or a year or two later if less vigorous. Some cultivars, in ideal conditions, will grow extremely quickly, so you'll need to keep a close eye on how they're doing. The idea is to strike a balance between doing too much too early, which will slow the tree down, and not doing enough to establish the shape you want.

You'll also need to check what policy your nursery follows. Many will supply trees with a strong central leader, and leave the decision about selecting platform branches to you. Others may well have nipped off the central growing point – perhaps at 0.9 metre – to encourage the growth of laterals. A few may encourage you to clear the trunk very quickly, but as this can leave the trees looking like lollipops, it is probably best avoided.

The monocone shape is designed for intensive planting and machine harvesting. The tree, although tall, is relatively narrow, and so the fruit falls down into a smaller area, making it easier to pick up. Establishing the monocone involves keeping a strong central leader taped to a stake or wire, and removing any competing vertical growth from lateral branches (so-called water shoots). This encourages the growth of fruit-bearing laterals. Eventually, the tree will topped at between 4 and 5 metres, and

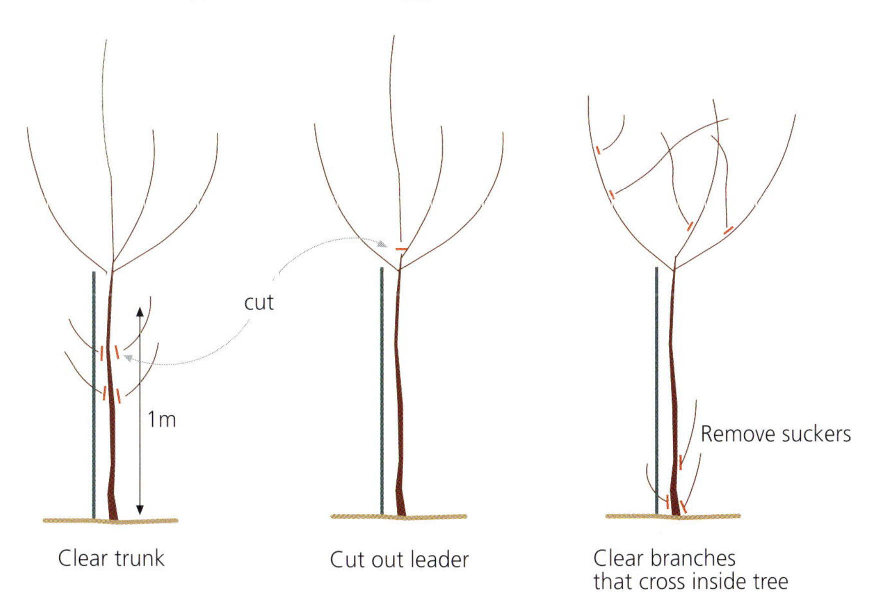

To establish the right shape, first clear the trunk, then cut out the vertical leader. In subsequent years, remove suckers and open up the vase shape.

may then be mechanically pruned (with a kind of vertical slasher or circular saw) on one side every second or third year. You'll still need to prune manually inside the tree, to remove vertical growth and to encourage light penetration.

Although the monocone is extensively discussed in the literature, and widely used in intensive plantings in Europe, especially Italy, some experts remain sceptical that the amount of pruning time required, which can be large, is justified by increased harvest efficiency.

The bushy vase is more suitable for hand picking than the monocone: the trees are kept to a more manageable height, and can be climbed into during harvest or pruning. But it makes sense to keep the tree to a single trunk with 1.2 metre clear at the base, just in case you decide to machine harvest at some time in the future. The vase itself is created by cutting out the central leader somewhere between 0.9 and 1.3 metres, at a point where there are at least three lateral branches to form the structure of the vase (and a comfortable height for climbing into the tree). Some experts recommend three branches, others as many as five – your choice will depend on what each tree has offer.

The basic shaping can start as early as the winter following the second growing season if the tree is growing really strongly, or as late as the fourth winter if it is less vigorous. Once the leader is cut out, the branches on the trunk below the new platform branches will also be cleared, though some experts recommend leaving the lower branches on for up to four or five years. You'll also need to remove any competing vertical water shoots and branches growing across inside the tree.

Whichever system you follow, you'll need to keep an eye open for suckers around the base of the tree. Left to grow, they will quickly take the tree back to a bush, and will sap energy from growth and fruiting in the main part of the tree. Sucker removal is possible with a contact herbicide, but only with extreme care, and not on young trees. The safest approach is to either manually rub off the buds as they form (hazelnut growers use a sort of chain-mail glove on their sucker-mad little beauties), long-handled secateurs (saves the back) or controlled grazing by sheep – but only when the trees are large enough for the main canopy to be out of reach. Sheep are not popular with some growers because their droppings can be roughly the same size and shape as olives, and difficult to sort out from the fruit at harvest! Nevertheless, they do return fertiliser to the trees.

Left unchecked, suckers will grow like wildfire.

Once the structure of the tree is formed, the annual pruning will be aimed at maintaining the tree's shape – 'so that a bird can fly through it', as the Spanish say – and to ensure that plenty of light is allowed to reach the centre of the tree. Some experts suggest that if the tree is heading into an on year, it may make sense to reduce the amount of last season's growth 'on' the tree. This should reduce the fruit-set, and encourage more new wood growth for fruiting in the following season.

The two basic essentials of maintenance pruning are heading off, used to top off the tree or to limit the spread of branches, and thinning. Heading off should be kept to a minimum, because it encourages many new buds to grow immediately beneath the cut, and they will need a lot of thinning. Thinning simply aims to stop overcrowding of growth.

Once the trees are mature, after perhaps 15 to 20 years (or longer), it may become necessary to rejuvenate the canopy. This is done by removing each of the main platform branches in turn, selecting new branches to take their place. Over a period of years, the whole canopy will be renewed, and fruiting vigour encouraged. Using this strategy, it's possible to keep very old trees in economic production.

Pruning technique is fairly straightforward. Don't cut either too close to the trunk (which may remove the bud you need to grow to replace the

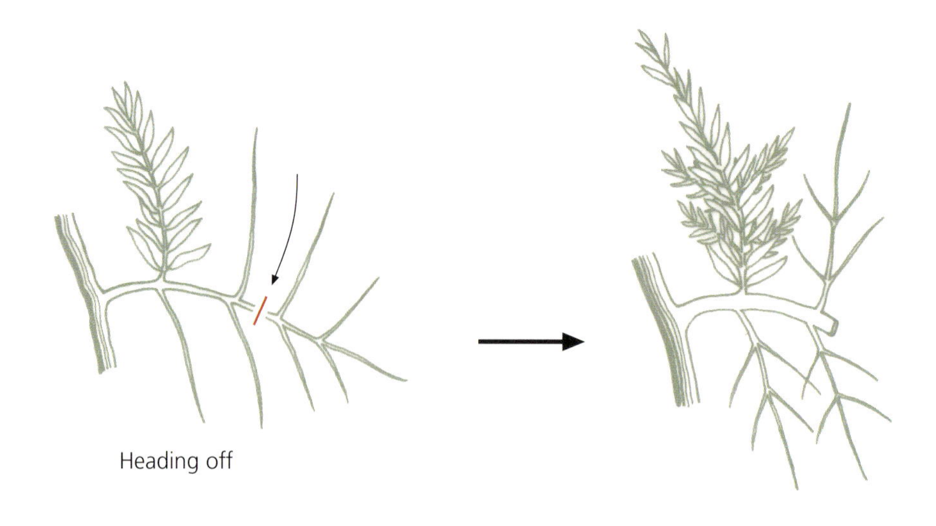

Heading off a branch stimulates the remaining side-shoots to develop, leading to a lot of vegetative growth.

branch) or too far along the branch (which will leave a stump that may rot, or sprout too much). Smaller branches can be removed with sharp secateurs or loppers. Major structural cuts will probably require a chain-saw. You can even get a rather dinky little compressed-air driven chain-saw that fits on the end of a long pole for reaching the top of the tree. Although small cuts can be left to heal themselves, very large cuts can be treated with a sealant paint, or any ordinary water-based paint. It is normal to sterilise the cutting blades of secateurs or loppers between trees (you carry a pot of disinfectant around with you), as this will help to reduce the risk of spreading infection within the grove. If your grove is clean, and New Zealand and Australia are free of most of the major olive diseases, then you may be able to get away without too much sterilising – perhaps doing it every few trees, or hours.

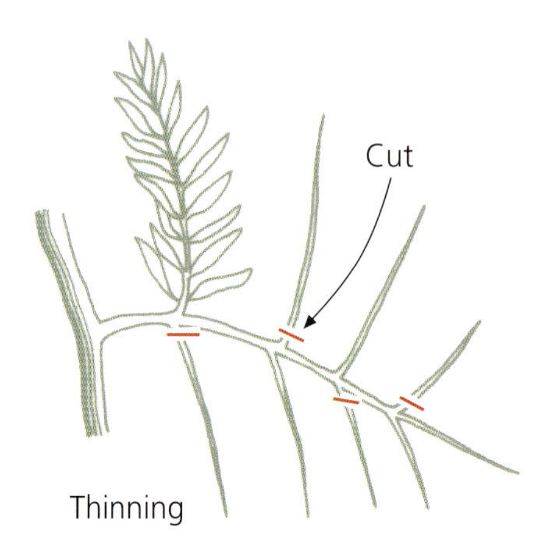

If branches become crowded, thinning allows more light into the tree and doesn't stimulate too much growth.

Right: Rejuvenating an old tree at Yakilo. Some of the old canopy has been left for pruning next year.

Below right: A year or two later, the new young canopy is developing rapidly and will need shaping.

Below: A slightly more extreme example: the chainsaw has simply lopped off all the old canopy.

Whatever your pruning strategy, you're likely to generate a considerable volume of trimmings. Wood from the largest pieces could be used for carving or turning – olive wood is in great demand in Europe, and could easily be used by crafty types. Olive-wood artefacts could even find a ready outlet through your farm shop or to the tourists being guided around your groves and press. The smaller stuff can either be burnt, as recommended by experts who are keen on reducing the risks of infections, parasites and so on, or put through a chipper or shredder and used as mulch around the trees.

Feeding the grove

Your original soil analysis will have shown if you have any specific nutrient and trace element

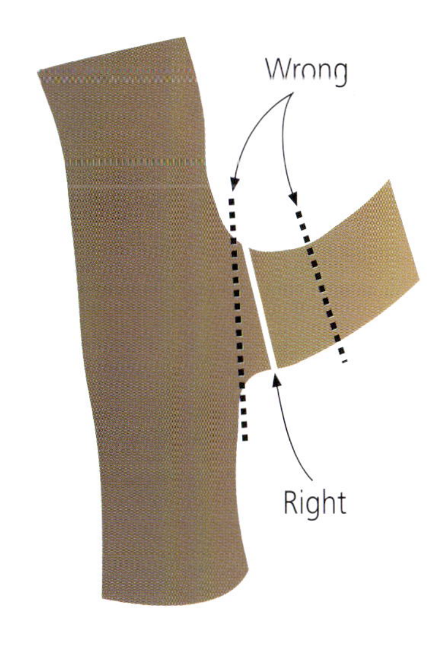

Pruning technique: don't cut close to the trunk, or too far along a branch.

deficiencies, and your soil treatment pre-planting should have been designed to address these. Once the trees are planted, the emphasis should be on encouraging them to grow vigorously. When they start to fruit, however, you should become more stingy with your feeding, as excessive fertiliser can cause the trees to favour growth over fruiting. Avoid overfertilising in any particular year, as this may push the trees into a pattern of alternate bearing.

The precise method of fertiliser delivery, and the mix of nutrients to be given, will vary enormously from grove to grove. It is impossible to give anything more than the most general of guidance, except to say that you should consult a local soil nutrition expert, and have leaf samples analysed on an annual basis – probably in mid-summer when the trees are working hardest and nutrient levels are at their lowest.

In the simplest of terms, you will have to put back into your trees what you take off them every year – that is, the nutritional equivalent of the fruit you pick, and the wood and leaves you prune. The carbon and hydrogen in these comes from the air and water; the other major elements of fertiliser required are nitrogen (N), phosphorus (P) and potassium (K). The most important of these is nitrogen (as it is with all plants), but P and K are also required for healthy growth. But the amounts and the ratio of these to be applied, and the methods used to apply them, will depend on the nutritional and physical properties of the soil, and this can be decided only by someone with an in-depth knowledge of your grove.

Trace elements, such as iron, manganese, zinc and boron are also important, and deficiencies can lead to various problems. The most problematic for New Zealand and Australian growers is boron deficiency, as soils in both countries are often very low in this element. Boron application is often necessary in commercial forestry in New Zealand, and although your grove may be lucky, it is certainly something you should monitor closely. Boron shortage shows itself in leaf yellowing, known as chlorosis, leaf-tip die-back or necrosis, and a distinctive deformity of the tip of the fruit known as 'monkey face'. Fortunately, it is easily corrected by foliar spraying, or by applying borax to the ground.

Organic growers will need to observe the same general principles, but will find other means of feeding the trees. I'm no expert in the field, but growing catch crops between the rows (if they are spaced widely enough) and then ploughing them back in will do wonders for the available nutrients and the soil structure. Similarly, applying compost around the trees as a mulch and feed will also be helpful, provided that the stuff is not too rich.

Irrigation

Once again, there is no easy rule of thumb for irrigating the mature grove. So much depends on the soil properties, the rainfall in any given season and the evapotranspiration from the grove, that no more than general guidelines can be given.

The key is to provide the tree with adequate moisture at the times when it is growing most strongly. Beyond that, a couple of strategies can be used to influence the characteristics of the fruit and to improve the harvest. The first is that a period of water shortage when the fruit is between 40 and 60 days old will stress the fruit in such a way that the size of the stone will be reduced. A smaller stone, and therefore a higher flesh-to-pit ratio, is important to table-olive growers. Later on, when the fruit is growing strongly – between 80 and 100 days old – giving the tree water will not only to increase the final size of the fruit, but will also increase the oil content.

There has been a lot of research into the relationship between irrigation and yield, and fruit characteristics, and many of the results are highly technical. There are useful summaries in the *IOOC Encyclopedia*, and the University of California Davis *Olive Grower's Handbook*. Of course, as in all things to do with growing olives in the Southern Hemisphere, it is not clear whether Northern Hemisphere experience will translate directly to our conditions, or whether other factors will complicate matters.

Coping with frost

If your grove is prone to frost, there are a number of things you can do to lessen the impact. Keeping the soil around the trees free of vegetation is helpful, as bare ground absorbs heat more readily during the day, and then radiates it away during the night, helping to warm the air. You should also keep the grass in the rows well mowed, as the longer the vegetation, the more air it will trap. The final safeguard, and this really applies only to small trees, is to spray them with a copper solution just before you expect frost. This shuts the tree down for a short while, and makes it better able to stand any damage from freezing.

Machinery

The amount of machinery you use in your grove will depend on the depth of your pockets and personal inclination, as well as the number of trees you have planted. Obviously, the larger the number of trees, the greater the need for mechanisation. In a very large grove, a four-wheel farm bike may be required just to allow you to keep an eye on the trees without having to walk several miles (unless you really want to).

If olives are a new venture on an existing farm, then much of the same machinery will be useful. The typical large farm tractor used for ploughing and haymaking will almost certainly be too large for use in the grove once the trees have become well established. A small orchard tractor will probably be more suitable, unless you need to provide power for a large sprayer for foliar feeding. Smaller and lifestyle growers will have to balance the amount of capital they need to invest against the usefulness of owning the gear. It may be more economic to use a local hire company than to buy a tractor upfront. Similarly, if you need to spray large mature trees, then using a contractor will make more sense than forking out tens of thousands of dollars to own the gear yourself.

Small growers will need a robust ride-on mower to keep the grass and weeds under control, or will have to hire a slasher-equipped tractor on a regular basis. A weed-eater may also be helpful, especially as the trees get larger. You will also need some means of spraying around the trees with herbicide, and perhaps foliar spraying the young trees. In a small grove, a backpack sprayer may be sufficient (but will be hard work). A larger-volume sprayer that can be mounted on a farm bike or the back of a tractor will make life a lot easier, and will be more useful as the trees get larger. You will also need some means of getting stuff into the grove, and fruit and prunings out; a trailer of some kind will be the best solution.

Only when growing olives on the largest scale is it economic to own your own mechanical harvesting machinery. Most growers use a local contractor, paying by the hour for their services. Manual harvesting can also be mechanised to a certain extent, with the use of compressed-air driven 'fingers' on the end of a long pole, but the basic requirements are a few cheap plastic hand rakes, a large amount of netting or shade cloth to spread around the tree, and plastic crates to take the fruit – the kind that stack on themselves, so that the fruit isn't damaged when being transported.

When pruning, all growers will need professional quality secateurs, long-handled loppers and a small chain-saw. Keeping them all sharp and clean will make life a lot easier.

PESTS, DISEASES AND OTHER PROBLEMS

Olives may be tough and hardy trees, but they are neither immune to disease, nor free from pests. They can also suffer from shortages of important trace elements, such as boron. Like any tree crop, they need a bit of tender loving care if they are to perform at their best. It is important that you wander through your grove on a regular basis, keeping an eye open for problems. Caught early, they will be much more easily solved.

Pests

Australia and New Zealand growers are lucky. The two worst pests that threaten olive production – the olive fruit fly, *Bactrocera oleae*, and the olive moth, *Prays oleae* – are confined to the Mediterranean, but the third, the black scale insect, *Saissetia oleae*, is present in both countries. It is thought that this particular scale insect originated from South Africa, where a complex of over 50 natural enemies help to keep populations in check. In Europe, there are fewer ecological checks and so the insect poses more of a problem, and the same is probably true in Australia and New Zealand.

Like all scale insects, this lives on the sap of the tree, extracting it through the bark and leaves, and exudes a sweet sticky honeydew. This encourages the growth of fungi, which form a black crust over the bark and vegetation. Severe infestations can damage trees badly.

The problem is easy to spot. Trees start to develop a black crust on limbs and branches, and little flattened insects can be seen clinging to the plant. Left untreated, it can be a serious threat to the vigour of the tree and, ultimately, the rest of the grove. Treatment is with a well-timed spray

Typical black scale infestation in a South Australian grove. The black crust is formed by a fungus living on the sticky exudations of the scale insect.

of white mineral oil – around mid-summer, when the youngest stages in the scale life cycle are on the move. This treatment is effective, and acceptable for organic growers, but it is possible to use insecticides instead. In the longer term it may be possible to introduce biological controls, but the economic importance of the business will have to increase significantly before that sort of effort will be regarded as worthwhile.

Australian pests

Australia's unique fauna, as well as the many introductions, provide a wide range of insects and other animals that can cause substantial damage to olive trees.

The Mediterranean fruit fly, *Ceratitis capitata*, is present in several states, and has been recorded as affecting olive fruit. It has been a substantial problem for olive growers in parts of Spain, and so it is being carefully monitored by Australian growers.

The olive tree bug, *Froggattia olivinia*, is a native insect that lives on the underside of the leaves, sucking sap and causing little yellow spots on the upper surface. Only 2 millimetres long, it has lace-like wings (but is

Wallaby damage in an olive grove, New South Wales. The poor tree has taken quite a battering.

In the same grove: cockatoo damage to the foil trunk-protector. Sharp beaks can do a lot of damage.

not a lacewing) and can cause significant damage if not controlled. Systemic insecticides seem to be the only solution at this stage.

The curculio beetle or apple weevil, *Otiorhynchus cribricollis*, is a rapacious little beast that leaves the ground at night to climb up the plant and eat the leaves. It is normally a serious danger only to young trees, in which case you will need to spray the ground around the trunk with a suitable treatment. An alternative approach is to use sticky or slippery bands around the trunk, though this is said to be less successful.

Other insect pests that have been reported as causing problems in parts of Australia include the Rutherglen bug, *Nysius vinitor*, the wingless grasshopper, the African black beetle, the pointed snail and codling moth.

Kangaroos can be a problem in many areas. They like to eat olive bark, and males are reported to 'fight' young trees, tearing branches off. Damage can be substantial. Possums are also a potential problem, with their liking for the fresh young shoots, and rabbits and hares are both fond of the bark of young trees. Foxes are also said to take the fruit.

Birds are a substantial problem. Parrots, especially the sulphur-crested cockatoo, not only enjoy eating the fruit, but will also bite through growing stems and attack the trunk. Other birds enjoy the fruit, and will take a large share of the crop if not scared away. Starlings in particular are blamed for the spread of wild olives around many parts of the country; it is surprising how often young seedlings are found next to fenceposts and other favoured perches.

Keep in touch with local growers and your local olive association. They will often be able to suggest solutions, or at least things to try when and if pests pose problems.

New Zealand pests

New Zealand olive growers are fortunate in that they face fewer, or at least different pest problems. Black scale is certainly present, and can cause problems if not treated early enough, but there are no vicious birds such as the cockatoo.

Leaf-roller caterpillars can be a real nuisance. Eggs laid by several species of moth on new young leaves develop into smallish caterpillars that use a kind of silk to glue three or four leaves at the tip of a branch together. The caterpillars then proceed to munch their way through the tender growing tips. Left unchecked, they can cause substantial damage. The first

Leaf-roller damage on one of the author's little trees. The insect glues the leaves of a growing tip together and chews its way through the tender young shoots.

caterpillars appeared on my trees within a month of planting. In that case, control was easy – I just squeezed the leaves together, crushing the caterpillar inside. With larger trees, you may need to consider spraying with an insecticide.

The grass-grub beetle, *Costelytra zealandica*, emerges in November and enjoys nothing more than chewing fresh young leaves. In severe cases small trees can lose lots of leaves and suffer a real setback. Grass grub can be controlled with special soil insect poisons, or with a biological soil drench.

Rabbits can pose a considerable problem for any planting of new young trees. They love to bite the soft bark, and will ringbark and kill unprotected trees. They will also chew their way through any leaves they can reach, and will do this even when there is plenty of fresh green grass in the vicinity. In any area where rabbits are present you will need to protect the trees with some form of tree guard – either tree boxes or tubes (preferably with a fairly large cross-section so that the young tree is not too tightly squeezed, and air can move around), or protective plastic sleeves round the bottom of the tree. There are also a number of protective sprays available, some of which are waterproof, that deter rabbits. I have heard good reports of Plant Skydd (and used it in my garden), but I prefer boxes for my trees – there's a certainty about them (and my trusty .22).

In New Zealand, possums are causing severe damage to large tracts of native forest. I have not yet heard of their attacking olives, but suspect it's only a matter of time before they will become a problem. Poisoning, trapping and shooting programmes are the only practical solution. Deer have also been known to eat the foliage, and this can be a problem with smaller trees..

New Zealand's parrots pose no significant threat to olive trees, but there are certainly plenty of starlings, blackbirds and thrushes to attack your crop, particularly as fruit ripens and loses some of its bitterness.

Diseases

The most widespread and damaging disease of olive trees is olive knot, which is endemic throughout the Mediterranean, and is said to be present in all olive-growing countries. It is caused by an infection with a *Pseudomonas* bacterium. The bark develops a knot-like canker up to 5 centimetres in size and bad infections can kill a lot of productive wood. The bacterium is spread by rain, and enters the tree through pruning cuts or damaged bark.

Australian growers have not reported any problems with olive knot, but in New Zealand the situation is a little more complicated. A similar (but not identical) bacterium causes a canker in native Oleaceae, and has caused infections in at least one grove. The Mediterranean olive knot has also arrived in the country on trees imported from an Italian nursery. These were destroyed before there was much opportunity for the disease to spread, but New Zealand growers are being urged to keep a close eye on their trees and report any suspicious cankers to the New Zealand Olive Association and the Ministry of Agriculture and Food (MAF).

In Europe, the problem is managed by a combination of cutting out affected wood and burning it, together with careful attention to hygiene while pruning. In Australia and New Zealand, infected trees are likely to be dug up and burnt.

A common problem, and one found in both Australia and New Zealand is peacock spot, sometimes known as peacock's eye. This fungus causes little black spots to appear on the leaves, and as these get larger they turn yellow. Fruit may develop a dusty-looking coating. Badly affected trees may suffer significant leaf-fall. The treatment is to spray the affected trees with a copper-based fungicide (such as Bordeaux mixture) in autumn and winter, after rain. Systemic fungicides have also been used to good effect.

As with all fungal problems, peacock spot is likely to be more of a problem in wetter or humid regions than in very dry areas. Some cultivars, including Leccino, Frantoio, Barouni, Sevillano, Ascolano and Nevadillo, are less prone to the disease.

Phytophthora, or root rot, is another fungal problem, usually caused by excessive moisture in the soil, and made worse if the root system has been damaged in some way, perhaps by wind rock or by ploughing or tilling too close to the tree. It can be treated by fungicides, but if it is turning up regularly it is a symptom that the site is wet, and will need remedial drainage, or simply too wet for a successful olive grove.

The list of nasty fungi gets longer: Verticillium wilt is thought to be rare in Australia and New Zealand, but is common in olive groves in California. Chunks of the canopy of mature trees can turn yellow and die, particularly in spells of hot weather, as a fungus infects the roots. The fungus is associated with soils that have previously been used for growing members of the potato family (Solanaceae) such as tomatoes, potatoes and tobacco.

One final fungus has been reported from Australia – anthracnose causes a deformity of the end of the fruit, which presents as a squashed, turned-in look to the tip of the fruit. This is called monkey face. It responds to copper fungicides.

Deficiencies

Nutrient and mineral deficiencies can express themselves in a number of ways. A shortage of boron, for example, can reduce vigour in general growth, and cause deformities in the fruit – another source of monkey face. Leaves can also turn yellow at the tips, and in extreme cases whole branches can die. New Zealand soils are often short of boron, and Barnea has been found to be particularly sensitive to a lack of this substance. Boron spraying is routine treatment for many tree crops in New Zealand, so should be easily arranged.

Leaf chlorosis, a generalised yellowing of the leaf, can be a signal of several kinds of deficiency, and so the best policy is to make sure that you give your trees and soil an annual check-up.

Other problems

Viral diseases certainly occur in olives, but at this stage of the development of the business in Australia and New Zealand little is known about which, if any, viruses are present. Soil nematodes may also pose problems in some areas, though this is not widespread, and in Australia parasites such as mistletoe and bridal veil (*Cassytha pubescens*) may affect old trees that have been untended for a long time. Old Man's Beard could pose similar threat in parts of New Zealand. Neither should be allowed to get anywhere near newly planted and properly managed groves.

OIL PRODUCTION

Much of harvesting olives for oil production is common sense, but there are guidelines to follow, some suggested by the IOOC, and some established by successful Southern Hemisphere producers of quality oil.

A lot will depend on the size of your operation and the market you're focusing on. Smaller growers aiming to produce a little high-quality oil for a boutique or gourmet market will have a very different set of priorities from the grower aiming to produce large quantities of oil at the lowest cost. The boutique grower may be prepared to hand-pick individual varieties as they ripen, perhaps picking through trees twice to get only fruit at the correct ripeness. Others will simply want to get the fruit off the trees as quickly as possible, at a time when the balance of ripeness in the crop gives the best trade-off between acidity and flavour.

Wherever you are on this continuum, one thing is clear: if you do things properly, you will almost certainly produce extra virgin olive oil. Even the largest broadacre operations in Australia should produce oil with an acidity low enough to easily meet the IOOC's 1 percent acidity qualification.

Ripeness

Researchers have shown that oil starts to accumulate in the fruit when it's about 80 days old, and that oil content continue to rise until the fruit finishes changing colour from green to red and black, when it begins to level off. If the tree is stressed by lack of water or extreme heat, this straightforward process might be interrupted, and oil levels reduced. Olive scientists therefore suggest that the best time to harvest is when the colour changes in the fruit have finished; this will give the best compromise between oil yields and flavour.

Ripe olives ready for the press. When the fruit has finished changing from green to black, oil content is at its maximum. *IOOC*

That nice simple picture is complicated by a number of factors. The first is that not all the fruit on any given tree will ripen at the same time. You'll have to make a judgement about what proportion of black fruit gives the best result in your trees. Another factor is that ripening patterns and their relationship to flavour vary from cultivar to cultivar, and season to season. In some cultivars, fruit left too long on the tree may develop undesirable flavours. And the longer the fruit is on the tree, the longer it is at risk from by insects and birds.

Another trade-off to consider is that some of the key chemicals in olive oil, the tocopherols and polyphenols thought to be significant players in the flavour and health benefits of the best oils, begin to diminish after (and even slightly before) the fruit finally becomes black.

It should be obvious from all this that setting a precise date for harvest, or even a precise formula for getting the best economic result, is going to be impossible. While grape growers can measure the sugar levels in their ripening fruit, and assess the flavours directly in the mouth, olive growers haven't yet found a simple indicator they can use. Even biting thoroughly ripe olives is a fairly unpleasant task. Some statistical methods can be used, which involve taking samples at random from a tree and then assessing the ripeness of the individual fruits, assigning them a score. When the scores are added up and fiddled with, they produce a maturation index. When this reaches the right level, you harvest the lot. In *Australian Olives* Michael Burr gives the details of the method used in Jaen in Spain. If it works for them, it could work for you, perhaps with adjustments to compensate for the cultivars you have in the ground, and the kind of oil you like to produce.

New growers will have to experiment. I plan to try pressing small quantities of fruit from different cultivars at different stages of ripeness, assessing the taste (and perhaps yield) until I have an idea about how to go about making my kind of oil. This is analogous to winemakers selecting batches of grapes for a particular wine, and blending wines to create the result they desire. As years go by, I hope to develop enough experience to achieve a consistent and tasty product.

In the grove

Olive oil quality depends on producing the best possible fruit. If the trees are healthy and well looked after, then the fruit has the best chance of being high quality – and yielding a high-quality oil. The more care you take of the trees and their precious cargo, the better your oil is likely to be.

The IOOC suggests that fruit should be picked from the tree, either manually or by tree shaker, instead of being allowed to drop naturally into nets over a long period. The olives themselves should be harvested at the 'right stage of ripeness' (not, unfortunately, defined), and growers should avoid picking fruit that is overgreen or overripe. If it's too green, yields will be low and the oil will have a leafy taste, while fruit that's too ripe will have 'weak organoleptic characteristics' (*translation*: won't taste particularly good).

Groves that are too small to justify using a mechanical harvesting machine, or where the owners are traditionally minded, will be hand-picked. There are a variety of mechanical aids, including pneumatically driven combs and shakers that can reach tall branches, but the simplest tool will be the most used – a modest plastic rake that you pull through the foliage, dislodging the fruit as you go. Ladders will also be essential as the trees get bigger. The fruit is usually allowed to fall onto nets spread on the ground, but table olive growers might pick into baskets or bags to keep fruit damage to a minimum. From the nets, the fruit is either put directly into plastic crates for transport, or washed first.

Organisation and teamwork are essential if the work is to be completed in a reasonable length of time. As a rule of thumb, teams of three or four people ought to be able to deal with a mature tree in around 10 minutes – but much will depend on the size of the tree, how well it is pruned and how much fruit it is carrying.

Mechanical harvesting of large, flat groves is likely to be the most economic option. This process is somewhat controversial. Some people consider that it may damage the tree in the long term, but the balance of opinion is that well-designed modern equipment will do little harm. The trunk is gripped by a large pair of jaws, and the whole tree is shaken until the fruit falls out, either into a integrated catcher, or onto nets on the ground. Large Israeli machines, working in pairs, are capable (it is claimed) of shaking three trees per minute. There are many types of machine, some with integrated catchers, and soon there will be even more.

The method does have its drawbacks. Cultivars differ in how firmly the fruit is attached to the tree, and in some cases a pre-harvest spray with a chemical designed to loosen the fruit may be necessary. Sprays have to be used with great care, or the trees may lose leaves as well as fruit. Even in ideal circumstances, the shakers will not remove all the fruit, and if a lot is left, you may need to go through and hand-pick the remainder.

Once picked, the olives should be delivered to the press as quickly as possible. If delay is unavoidable, they should be stored in shallow layers in a cool place. Don't use traditional jute sacks, because the fruit can begin to heat up and ferment, encouraging fungal growth. Self-stacking plastic crates are ideal, though if they are too large the fruit at the bottom may bruise. If any of the olives are badly bruised, or otherwise unsound, they should not be stored, but processed for oil straight away. Any olives that are gathered off the ground should be washed, to avoid dirt contaminating the olive paste.

Many of the more fastidious Australian growers have begun to wash the olives in the grove as soon as they are picked. One has even stored his fruit under nitrogen while waiting for it reach the press, to slow down the rate of oxidation, and therefore deterioration. In some of the more remote areas, or places far from a press, growers use refrigerated transport to get the fruit to the press in the best condition.

At the press

The IOOC suggests that any olive press or mill should be very clean and hygienic, and that growers should be especially careful to avoid poor-quality olives contaminating or tainting the process. The best method of crushing the olives is with a stone mill. This is preferable to a hammer mill, the IOOC says, because it does a better job of pulping the fruit and doesn't emulsify the oil or heat the paste. Mixing of the paste should be done for only a short time, and the temperature should not be allowed to rise above 25 to 30°C, though it is permissible to increase both slightly if the paste is 'unmanageable' – presumably too stiff.

If the press is a traditional hydraulic device, the mats that hold the paste must be scrupulously clean. Paste left clinging to the mats may begin to go off very quickly in warm weather, or if stored overnight. Some presses now store the mats in freezers during the pressing season to ensure that this can't happen. There have even been suggestions that inadequately cleaned mats may contaminate the following season's oil, so it is important to those who press your fruit to understand the importance of cleaning their mats. If you want to keep control in your hands, but can't afford your own press, a compromise may be to invest in a set of mats.

The guidelines for centrifuge oil extraction are slightly more complex, but are aimed at minimising the amount of extra water added to the olive paste and restricting its temperature to no more than 25°C. Separating the oil from the mixture should be done as soon as possible, and extra water should not be used, as 'washing' the oil at this stage can spoil its keeping properties.

Once pressed, the oil should be stored in what the IOOC terms 'appropriate containers' – in practice, this means in stainless steel tanks of various kinds. Filtering, if you decide it's necessary, should be done before long-term storage. Unfiltered oil should be allowed to settle in tanks designed to allow the dregs to be tapped off. In either case, you're aiming to remove the dregs and bits of fruit before they can begin to go off and taint the oil. Long-term storage tanks should protect the oil from light, heat and exposure to air.

This sort of advice might seem obvious to people used to the cleanliness and hygiene required by winemaking or dairying, but the IOOC guidelines were not drawn up with the Southern Hemisphere in mind. They were designed to help the Mediterranean industry establish some standards, and to replace arcane traditions and dubious procedures with techniques that have been shown to work. We ignore them at our peril!

Choosing a press

Growers who want to invest in their own pressing plant will be able to control the process from start to finish, but the capital investment required is high and most growers will either sell their fruit, or contract the pressing to a local mill. Choosing who will press your fruit might seem a distant luxury to growers who have no press within easy reach but, as the business expands, presses are likely to be established wherever there is enough fruit to justify the capital expenditure. Even if choice is restricted by circumstance, some basic factors still have to be considered.

Location is important: the further away the press, the greater the transport cost, both in terms of the distance and the need to keep the fruit in ideal, perhaps even refrigerated, conditions. The type of machinery used in the press is also important, especially if you have formed an opinion about which is best for your oil.

Presses fall into three broad categories: small all-in-one machines that combine a hammer mill with a small centrifuge and that can process up to

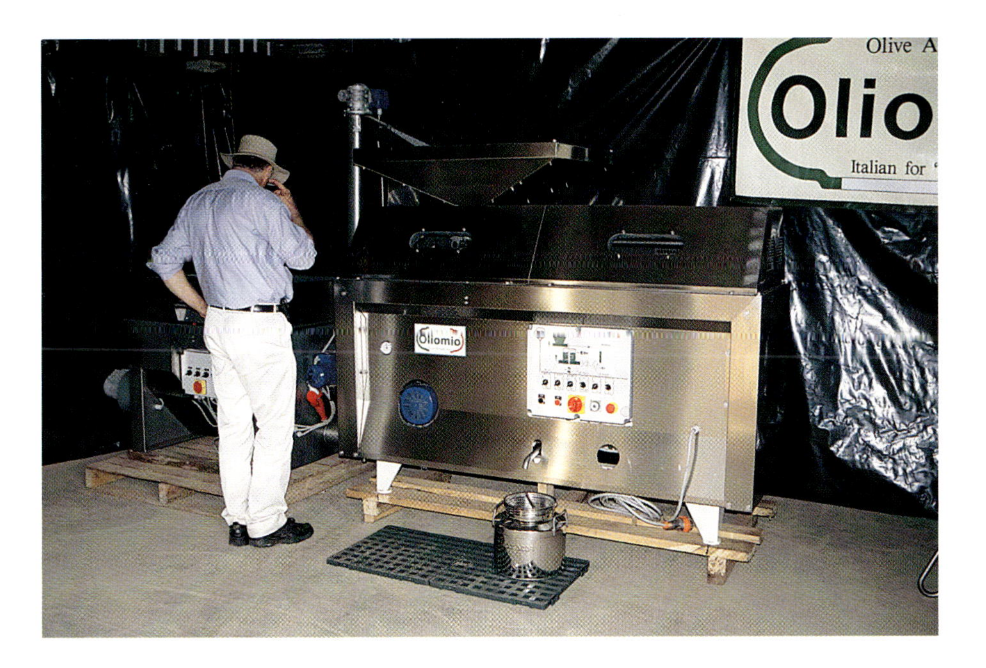

An Oliomio continuous-process machine. Olives go into the washer at the left, then into the hopper on top. Oil comes out at the front.

100 kilograms per hour, larger traditional hydraulic presses that can handle upwards of 100 kilograms per hour; and very large continuous process centrifugal mills capable of handling several tonnes of fruit per hour.

A relatively small hydraulic press, capable of pressing 100 kilograms of paste per hour. A white nylon pressing mat can be seen at the base.

The smallest presses, exemplified by the Oliomio machines, are proving popular with small growers who have some capital to spare. The largest claims to handle 100 kilograms per hour in a continuous process, and is said to produce good oil. You're most likely to encounter these machines in small operations – perhaps neighbours who want to recover their capital investment by contract-pressing local fruit.

Hydraulic presses are a much bigger investment, and so are likely to be associated with a larger grower, a group of growers or a co-operative. Traditionalists will probably favour this sort of plant, on the grounds that their gourmet single-estate extra virgin can be produced only by being squeezed between mats. On the other hand, hygiene and cleanliness will be very important factors to consider. If your wonderful fruit is contaminated by pressing in dirty mats, your oil will lose a lot of its value.

Large centrifuge plants will be the realm of the bigger grower co-operatives or oil brands. They will be established only in areas where the owners can reasonably expect to make a return on their investment, and that means where there is a lot of fruit to process. If you have a large grove, producing hundreds of tonnes of fruit, you will naturally gravitate to this sort of establishment.

Smaller growers have a wider choice, but once again size will be important. If you have 200 trees producing 35 kilograms fruit per tree, you will have 7000 kilograms or 7 tonnes of fruit to deal with. On a small machine, handling perhaps 100 kilograms per hour, your fruit will take 70 hours to process – seven 10-hour days, or nearly three days if the machine is working 24 hours a day.

If it takes 15 minutes for a team of four people to hand-pick each tree, you'll be producing 140 kilograms per hour. If you pick for 10 hours a day, then every evening you'll have 1.4 tonnes of fruit to deal with. The ideal, as we've seen, is to press the fruit as soon as possible, so it would be nice to find a press that could handle that amount overnight.

Whatever the size of your grove, you'll have to match the press capacity with your production, to avoid having to store fruit for long periods – or the expense of storing it properly. In busy periods, good presses will inevitably become congested, so you'll probably need to book a slot.

There is another, more controversial, point to consider when using a press. In any continuous-flow process, your fruit is inevitably going to be mixed with the remnants of the preceding batch, and the batch that follows.

Once you have the oil, you have to choose a bottle. A wide variety of colours, shapes and sizes are available. *Kirk Hargreaves*

Where does your oil start, and your neighbour's finish? That may not matter if you're selling your fruit to a large outfit or co-operative, but it does if you want to market your oil under your own label. The same is true, if a little less obviously, in a traditional press. The mats may contain bits of somebody else's fruit, or the crusher and mixer may have paste remaining in corners. All the more reason to insist on scrupulous cleanliness.

Bottling and marketing

Getting the oil into bottles may seem a fairly straightforward sort of task, but choosing the bottle and designing a label certainly isn't. Small oil producers will probably settle for a small unmechanised bottling and sealing system; large presses will certainly justify an automated system. There may even be a mobile bottling plant available, which can turn up at your farm and bottle all your oil in a few hours or days.

Bottles come in a huge variety of shapes and colours, not to mention sizes, and the choice you make will be an integral part of the marketing of your oil. The label – the combination of words, images and legally necessary statements – is also a key part of that marketing strategy. How you arrive at that strategy is well beyond the scope of this book, but your objective is to put your Product in the right Place, at the right Price and to then Promote it – the four P's of marketing, in other words. Define your goals, determine a strategy and then mind your P's (and wait for the queues).

There can be many pitfalls in marketing, some not at all obvious. It's a trivial example, but what about the highly acclaimed, very expensive, prize-winning oil from Australia that comes in a bottle so tall it won't fit into my kitchen cupboards? It irritates me every time I have to get it out of the broom cupboard. It probably doesn't fit many retail shelving systems, either. Damn fine oil, though, which is why they get away with it.

It isn't enough to make good oil. You may produce a few thousand litres of the finest oil in the world, but if it's not in the right bottle with a good label, stocked by the right kind of shop at the right price, and if it's not being drawn to the attention of the right kind of buyer by good promotion and advertising, it won't sell. The more product you produce, the more critical your marketing. If your oil is going to slug it out on supermarket shelves with the big brands, you'll need to play the game by their rules, and that means being relentlessly professional in your marketing.

OLIVES FOR EATING

There are many ways of treating olives to get rid of the oleuropein that makes their flesh so bitter. Large commercial operations use scientifically controlled and hygienically clean processes to make sure that their products are safe to eat. Small-scale producers may be a little more cavalier, particularly if the olives are for private use, but there are still many different recipes to choose from.

From an olive grower's point of view, table olives can either be sold fresh in the market, for consumers who like to pickle their own, or sold to a major olive processor. In Australia, the sale of fresh olives can be reasonably lucrative, as all the major cities have substantial populations of people with Mediterranean cultural backgrounds. The New Zealand grower is less fortunate, though it is possible that a fresh market might emerge.

Growers who aim to sell their fruit to a large processor will naturally want to work closely with the buyer to ensure that they produce fruit of the right quality to attract the best price, and their interest in the process used to bring them to table is likely to be theoretical, at best. On the other hand, it would be a strange grower who didn't want to pickle a few olives from his own trees, if only to feed to guests. Recipes are given at the end of this chapter.

In the grove, the harvest of olives for the table is usually done at one of three stages, depending on the desired end result. Green ripe olives can either be processed green, or turned black in processing (the Californian way). Olives that are halfway from green to black, sometimes called violets, and fully ripe black olives are usually left as nature intended. Some growers may leave certain olives (especially Kalamata) on the tree for a long time, until they begin to wrinkle. Table olives are normally hand picked, so that bruising can be kept to a minimum, and so that only fruit of the right degree of ripeness can be selected.

In the supermarket take a look at the varieties of olives available. On the shelves there will be Spanish green olives, some on the stone, some pitted, and others stuffed with bits of red pepper or onion, plus green and black Greek olives in brine, and probably several more besides. At the deli counter, there will be plain Kalamata olives, flavoured olives in various styles, and mixed olives of all sorts. Each style of olive demands a different process.

Whatever style you want to create, the main aim is to remove as much as possible of the oleuropein from the flesh, while leaving the fruit full of

Vast pickling tanks at Viva Olives' processing plant, Loxton, South Australia.

flavour and looking appetising. There are two principal methods: soaking the fruit in lye, a solution of caustic soda ($NaOH$), which is brutal but quick, or a longer process of soaking in brine.

OLIVE STYLES

Spanish-style green olives

The fruit is harvested as soon as it is green ripe, and before any colour change has started. After washing, the olives are put into 1.6 – 2.4 percent lye, and stay there for up to 15 hours until the lye has penetrated about two-thirds of the way through the flesh towards the stone. The olives are washed repeatedly to remove the lye, and then put into brine (at about 10 percent). If conditions are right – and the skill of the operator is important – then a fermentation begins, with bacteria changing sugars in the olive into lactic acid. This gives the olives their characteristic flavour and, if done properly, means that the olives will keep for a long time. After fermentation, the olives are sorted, pitted and stuffed if desired, and packed in brine for sale.

Picholine style

This method is used in the south of France, as well as Morocco and Algeria. It's essentially the same as the Spanish method, but after the lye treatment, no fermentation is allowed to take place. The olives go into a 3 percent brine, which is gradually increased to 10 percent. They're ready to eat in about 10 days.

California-style black olives

These are harvested just as the fruit are beginning to turn black, and given alternate treatments with 1-1.5 percent lye solutions and water. Compressed air is blown through the water baths, which turns the olives black. An iron compound (ferrous gluconate) is often used to complete the process, fixing the black colour. The olives are then pasteurised and packed in brine. They are not considered to have much flavour, but seem to go down well in the United States.

Greek-style black olives in brine

The fruit is picked when ripe – completely black – but before any shrivelling starts. It is then washed and put into an 8 percent brine solution, and kept under the surface away from air. A fermentation may start, and the brine concentration will drop, so extra salt will be added to keep the solution up at 8 percent or more. The time needed to remove the bitterness varies considerably, but once the fruit reaches the desired stage, which can take up to nine months, it is washed and packed in brine or vinegar for sale.

Kalamata-style olives in brine

This is a much quicker process. The olives are sliced on either side with a sharp knife, which makes it much easier for the oleuropein to be washed out. They are then soaked in water or brine, usually changed regularly, and will be ready in a week or two. They are then put into a vinegar solution for a couple of days, and finally packed in a fresh 2 percent brine, perhaps with a lemon or two.

Pickled olives ready for packing are sorted by hand at Viva Olives.

Black olives in salt

These are really delicious, but very salty. The olives are picked when they are overripe, and layered with dry salt in baskets or other containers that will allow the water given out by the olives to drain out. After 30-40 days the olives are ready, and are usually stored dry, perhaps in fresh salt. They don't keep very well, and so should be eaten fairly quickly. You won't have any trouble...

Flavourings

Olives are often packed with various flavourings – herbs, spices, orange peel, lemons. These are normally added after processing is complete, in the final packaging or storage container. If you're trying this at home, you will need to be careful that the things you put in are clean, and will not go off in the brine.

A simple home method

This is really just a home version of the Kalamata treatment described above. First, a caveat. Be very careful to observe simple food hygiene precautions: clean hands, sterilising jars, throwing out anything that grows mould and so on.

The olives should be black ripe, but you can experiment – black all the way through to the stone, or halfway, will depend on the variety and how you like the end result. Cut a slit on each side of the fruit with a sharp knife, and then put them into a bucket/tub of water with a plate or something similar to keep the olives below the surface; don't expose them to air. Change the water every day, and skim off any scum that may appear on the top. After about a week, the olives should have lost most of their bitterness, but this is something you can decide for yourself only by tasting them.

Once they're ready, prepare a brine by adding 100 to 120 grams of sea salt per litre of water, and bringing it to the boil. Let it cool, and then pour over the olives. Make sure the olives are kept beneath the surface, and layer the surface with oil, to exclude air. They can then be left for up to six months, but will probably be ready after three.

Bottle them in sterilised jars, in either a weaker brine (50 grams per litre), to which you can add various flavourings, or in vinegar – wine vinegar diluted three to one, plus flavourings, if you want. The olives should keep well, but throw out any that show signs of going off or that grow mouldy bits. You should also keep opened jars in the fridge.

OLIVE CUISINE

There are shelves full of books on cooking with olives and olive oil, and I'm not about to start competing with greats like Elizabeth David or Marcella Hazan. In this short section, I simply want to draw your attention to a few principles, and offer a few directions that may inspire, by drawing heavily on the cuisines of the Mediterranean. After all, real olive growers should eat the stuff – and the more the better.

Broadly speaking, olive oil can be used in three ways: as a flavouring in its own right, for frying or in baking, and as a medium for other flavours. Table olives are usually used for their own assertive flavour contribution to salads, pizzas and the like, but I can't (and won't) avoid mentioning tapenade, the classic, earthy paste of black olives from the south of France.

Olive oil is used in all Mediterranean cuisines and essential for the concentrated energy it provides, and its flavour contribution. At its simplest, olive oil can be used as a dressing for salads and vegetables.

Dressings

It is sometimes said that Italians view salad as merely a convenient medium for tasting oil, and there's certainly nothing wrong with taking a few very fresh salad leaves — a mesclun mixture, perhaps — washing it, and then coating it in a little of your very best oil. Add nothing more, and eat it on its own, or perhaps as a salad course after the meat.

For a slightly more complicated salad dressing add some wine vinegar to the oil (never more than one-third vinegar, in my view) together with some salt and pepper, and whisk it all together. Too much vinegar will swamp the taste of the oil, so err on the side of caution. A simple vinaigrette like this can be a thing of real beauty if made with a good balsamic vinegar and fine extra virgin oil.

Pinzimonio

In this classic Tuscan dish (called *cazzimperio* in Rome) olive oil is the star. When you go to a restaurant, a basket of tender spring vegetables are brought to your table to welcome you – little spring onions, young fennel bulbs, baby carrots, small celery stalks, even tender baby artichokes – together with a bowl of good oil, to which has been added a little salt and pepper. You simply dip the vegetables in the oil, and enjoy the ultimate healthy starter.

Steamed vegetables (very fresh vegetables that have been cooked quickly) take on a whole new life when dressed with a suitable oil. You don't need much, a few tablespoons will do: just enough to coat the vegetables and make them glisten. They will not only look great on the plate, but will taste better as well.

Another approach is to use good oil as a sort of dressing for soup. You just add a generous dollop to the soup as you serve it, leaving it up to the guest to stir it in as desired. This is particularly effective with simple summer vegetable soups, or a cold soup like gazpacho, but can also be an integral part of much heartier bean soup/stews such as *pasta e fagioli* and *ribollita*.

Fettunta (bruschetta)

The classic way of tasting oil remains the Italian habit of pouring it on bread. Long before restaurants replaced the butter dish with a little bowl of oil for you to dip your bread in, Tuscans were enjoying their *fettunta* (now better known worldwide as *bruschetta*).

Cut a thick slice of good day-old bread, preferably Italian country-style bread such as *pugliese* or, better still, the classic Tuscan *pane sciocco*, and grill it on both sides until it is just beginning to char. Take a clove of garlic and rub it over the hot bread, so that the garlic melts into the surface of the bread. Then pour over it the best olive oil you've got, and tuck in.

A slightly more adventurous version can be made using tomato – but only with the ripest tomatoes you can find. After rubbing in the garlic, slice a tomato in half and rub it into the surface of the bread so that all the juices are mopped up. Then pour the oil on top and, if you've got some fresh basil, add a leaf or two as well.

If all the ingredients are right, and the sun is shining, the first mouthful can transport me to a Tuscan hillside on a summer's afternoon.

Salmoriglio

Olive oil is also the basis for many sauces for fish and meat, especially such things as *salmoriglio*, where the pungent flavours of fresh herbs such as oregano, thyme or parsley combine with the oil and a little lemon juice to create something that can transform a simple boiled or grilled fish.

The ratio of lemon juice to oil is normally about 1 to 4. If you're using parsley, you just chop it up and add it to the oil, but oregano and thyme should be pounded in a pestle and mortar with a little salt. Be generous with the herbs, but be warned that the flavours can be very powerful.

A slightly more chic version would drop the lemon juice, and add finely chopped fresh red chilli and garlic to the parsley – brilliant on almost anything (especially grilled squid).

Sauce for pasta

A very similar sort of dressing or sauce can be made for pasta, either for a very quick and tasty snack or starter, or as a component of a larger main course.

Put a good slurp of assertive oil, something with bags of flavour, in a small pan. Add some chopped garlic (as many cloves as you can stand), and heat the oil to the point where the garlic is just beginning to fry (if you let it continue frying, it will turn brown and bitter.)

Take it off the heat, and add chopped parsley. Stir it all together; you don't want it to be too thick, so add more oil if necessary.

Drain your pasta – I use spaghetti, but anything long will do – and quickly stir in the sauce so that every strand is coated. Serve immediately with good freshly grated *parmesan* or *pecorino*. I also serve this as the base for a pan-fried chicken breast (leaving out the cheese).

Marinade

Olive oil is excellent as a base for a marinade, combining well with the flavours of meat such as lamb and herbs such as rosemary. My own favourite is simplicity itself: take a boned (butterfly) leg of lamb and leave it to marinate for a few hours in olive oil to which you've added a little white wine vinegar (balsamic's good too), a handful of chopped rosemary, some chopped garlic and some freshly ground black pepper. Grill the meat, preferably on a barbecue, to your preferred degree of pinkness.

Puree de Pommes de Terre a l'Huile de'Olive

Now that we're into the world of warming up oil, rather than using it straight from the bottle, I feel you should be acquainted with what many people regard as the ultimate mashed potato. Devised by French chef Freddy Girardet, it became a cult dish in London, and has been much imitated around the world. It is not for the faint-hearted.

Boil 275 grams of peeled or scraped new potatoes until they are tender enough to mash. Drain them well, and either push them through a sieve or mill them through a *mouli legumes*.

Add 150 millilitres of warmed double cream, and beat the lot with a whisk until the cream has been absorbed.

Then add 150 millilitres of a fruity olive oil and carry on whisking. Keep going until the oil has gone, and then carry on for a while longer. The more you beat, the lighter and fluffier it will be.

Season with salt, freshly ground black pepper and a little cayenne pepper. Luxuriate.

Frying

Olive oil is a perfect oil for frying of all kinds – it can be heated to a higher temperature than most oils and can be reused many times before becoming rancid. One day, perhaps, fast-food outlets will fry their chips in olive oil, and our children's diet will improve out of all recognition (as will the flavour of their chips).

The Italians have a great range of fried dishes, from the classic *fritto misto* of the day's catch at the Adriatic harbourside, through deep-fried stuffed zucchini flowers to pastries fried in oil in the Spanish manner.

A rather aristocratic alternative to chips

Take a large heavy frying pan and cover the bottom in good extra virgin oil. You need to put in enough to make the oil come halfway up the pieces of potato – only practice will tell you how much. Peel enough potatoes to feed your family, and then chop them into cubes roughly 1 centimetre square.

Heat the oil until it's just beginning to smoke, and then add the potato cubes, spreading them around so that they form a single layer.

While they're frying, pick a handful of rosemary and strip the leaves off the stalk. Separate some cloves of garlic from a head – say two or three per person – but leave them unpeeled.

When the potato cubes have started to brown on the underside, give them a stir and add the rosemary and garlic, a generous twist of black pepper from the mill, and a sprinkle of sea salt.

Carry on frying and stirring until the cubes are a crispy golden brown all over, then drain them on kitchen paper and serve as hot as possible, garlic, fried rosemary and all.

This is immensely popular with my family, so much so that, over the years, our frying pans have got steadily larger. Another version leaves out the garlic and rosemary, but substitutes sage leaves – which become crisp and tasty and lend the potatoes a lovely flavour.

Baking

Olive oil can also be used in baking. Many Mediterranean breads feature olive oil – it gives them a richness and depth of flavour that can't be matched – but olive oil can also be used in almost any baking, from pastry dough to sponges and muffins: anywhere where you might use butter, margarine or other cooking fats.

If you're baking something delicate, you might not want the full assertive flavour of a fruity green extra virgin, so there is a legitimate use for the so-called 'light' oils (which are light only in flavour).

You may need to experiment a little to get the right quantities of oil to substitute, but the following guidelines should help:

50 grams butter or margarine = 40 millilitres olive oil
115 grams butter or margarine = 90 millilitres olive oil
225 grams butter or margarine = 165 millilitres olive oil

For anything that calls for less than 50 grams do a simple one-for-one conversion (grams to millilitres) to avoid the risk that the baking might turn out tough.

Olive oil is a wonderful vehicle for other flavours. At many French and Italian pizzerias, every table has its own bottle of olive oil, stuffed with chillies and herbs, ready to spice up the pizzas with an often fiery bite.

Infused oils

Infused oils are available in all sorts of flavours, from white truffle through to lemon and any kind of herb. Truffle oil is very expensive (if you can find it), but it is the only way of getting an idea of the real flavour of this expensive delicacy without travelling to Italy. The pronounced musty, mushroomy flavour is weakened in cooking so the oil should be added, a few drops at a time, to salads, and as a finishing touch to pasta or risotto.

Making your own infused oils is pretty straightforward, though you do need to be careful that the things you steep in the oil are clean and don't go off. Remove them after a couple of weeks, or when the oil has got the intensity of flavour you're after.

A few fresh chillies, with shallow cuts scored along the sides, will give any oil a kick after a couple of weeks. You can also use dried chillies – the little red ones are very effective. Dried wild mushrooms, especially *Boletus edulis* – the *cèpe de Bordeaux*, or *porcini* – will give the oil a marvellous smoky richness. Garlic is an obvious choice, basil is traditional (especially in the famous Genoese *pesto*) and the only limitation is your own imagination.

Pickled or table olives

The range of flavours available in pickled or table olives is vast, from the salty richness of Greek dry-cured Kalamatas to the velvety subtlety of a black olive marinated in oil and spices. There are literally thousands of ways of treating and flavouring olives, far too many to go into here, so I'll confine myself to giving you *tapenade*, the classic black olive paste from Provence. This version is from Elizabeth David's classic *French Provincial Cooking*.

You need 24 stoned black olives, 8 anchovy fillets (salted are better than oiled), 2 heaped tablespoons of capers, 60 grams of flaked tuna, lemon juice and olive oil.

Put all the solid ingredients into a pestle and mortar and pound them until they're a thick puree. Alternatively, you could give them a quick whizz in a food processor.

Then add the olive oil – 'about a coffee-cupful, after-dinner size' says Elizabeth David – a little at a time, as if making mayonnaise, and finally add a squeeze of lemon juice. A drop of cognac could be added, and perhaps a little mustard, but no salt.

This can be eaten on toasted slices of French bread, or served with hard-boiled eggs, or even left on the table as a kind of dip.

Elizabeth David is also sometimes credited as being the first person to introduce Britain to the delights of single-estate extra virgin olive oil, when her shop in London's Soho imported and sold oil from Poggio Lamentano, a Tuscan estate owned and run by a Scottish woman and her Polish husband, a man called Aleksander Zyw. Elizabeth David became greatly attached to the oil, and long after selling her interest in the shop, continued to receive special shipments for her own use. Shortly before she died in 1992, she wrote: 'The horrors of the corn oil era have at last receded. Last year when Zyw oil ran out there were cries of distress and disbelief – well, it won't happen again.' You may never taste Poggio Lamentano, but you will surely come to agree with her sentiments.

BOOKS, CONTACTS, AND OTHER SOURCES

Technical books

Australian Olives, A Guide For Growers And Producers Of Virgin Oils, by Michael Burr, Third edition, January 1998; ISBN 0 646 348 159 (*Michael Burr, PO Box 142, Stepney, Adelaide 5069, Australia; betaburr@senet.com.au*) Very useful to any grower and an absolute must for any Australian grower. This self-published book contains the distilled wisdom of Australia's leading olive guru, and is interesting, informative and amusing.

The Good Oil, Growing Olives In New Zealand, by Mike Ponder, First edition, 1998; ISBN 0 473 05448 5 (*Wenlock House, Blenheim, New Zealand*) An interesting and useful summary of twelve years' experience of growing and selling olives and their oil in New Zealand.

World Olive Encyclopaedia, First edition, 1996; ISBN 84 01 61881 9 (*International Olive Oil Council, Madrid*) Covers just about everything established growers might need to know, though the focus is very firmly on the Mediterranean. A lot of it is rather hard going, being translated from original semi-scientific articles in various languages. Very expensive.

Olive Production Manual, by Louise Ferguson, G. Steven Sibbett and George C. Martin, First edition, 1994; ISBN 1 879906 15 5 (*University of California, Division of Agriculture and Natural Resources, Oakland, California*) Useful, especially on producing California-style table olives.

An Introduction To Olive Oil Processing, From Picking To Pouring, by Julian Archer, First edition, 1997; ISBN 0 646 34515 X (*Olives Australia, 16 McGarva Rd, Grantham, Queensland 4347, Australia; oliveaus@ozemail.com.au*) A useful overview of olive-oil production.

Olive Oil And Health, by Publio Viola, 1996 (*IOOC, Madrid*) Very useful on olive oil chemistry and health.

General books

Olives, The Life And Lore Of A Noble Fruit, by Mort Rosenblum, First edition, 1996; ISBN 0 86547 503 2 (*North Point Press, New York*) An enthusiast's guide to the olive world. Full of intriguing fact and personal opinion – a great read.

Extra Virgin, An Australian Companion To Olives And Olive Oil, by Karen Reichelt with Michael Burr, First edition, 1997; ISBN 1 86254 417 4 (*Wakefield Press, Kent Town, South Australia*) Good on the history of olive growing in Australia, with a wide range of recipes from top Aussie restaurateurs.

Olives: The New Passion, by Genevieve Noser with Professor Don Beaven, First edition, 1997; ISBN 0 670 87932 0 (*Viking, Auckland*) A summary of the history and development of olive growing in New Zealand, with some interesting comments from current growers and a useful section on health issues.

The Olive Tree, The Olive, The Oil (*IOOC, Madrid*) An overview of the whole olive business, from history to recipes, but a bit technical in places.

The Olive Oil Companion, by Judy Ridgway, First edition, 1997; ISBN 0 670 87743 3 (*Viking, Ringwood, Australia*) After a brief roundup of the world of olive oils, dives into an estate-by- estate consideration of some of the world's best. Only one Australian and one New Zealand oil, but plenty of Tuscans.

The Goodness Of Olive Oil, by John Midgley, First edition 1992; ISBN 1 85145 995 2 (*Pavilion, London*) A slim volume. Gives a brief overview of the olive world, then lots of recipes, mostly good ones.

Journals

Olivae (5 issues per year) is the official magazine of the International Olive Oil Council, and covers IOOC doings, the world oil and table-olive trade, and publishes the latest research on olive growing.

The Olive Press (quarterly) is the newsletter of the Australian Olive Association. Includes lots of material on the business in Australia and useful articles on all aspects of olive growing.

Australian Olive Grower (5 per year) is a somewhat glossier guide to the Australian business, published by Olives Australia, the giant Queensland nursery.

The New Zealand Olive Association also publishes a newsletter.

Websites and the internet

These websites were all "live" at the time of going to press, but that's no guarantee that they'll be around when you read this. Use Internet search engines or one of the "metasearch" programs that ask lots of search engines the same question and then summarise the results.

http://www.australianolives.com.au/
The website for the Australian Olive Association. They have a very useful links page.

http://www.oliveaustralia.aust.com/
The big Queensland nursery's face to the wired world. Lots of very useful information, including their "Olifax" series of information sheets.

http://fruitsandnuts.ucdavis.edu/olive2.html
The University of California's Davis centre is the major non-European centre of olive research and this site has lots of useful information.

http://www.oliocarli.com/museo/index.htm
An Italian museum of olive oil. Full of fascinating stuff on the tree, its history and cultivation.

http://www.theolivepress.com/
This Californian company presses local extra virgins.

http://www.asoliva.com/
Home page for the Spanish oil exporters' association.

http://www.oliveoil.net/
A useful site: full of links to web sites around the world

Contacts

International Olive Oil Council
Principe de Vergara, 154, 28002 Madrid, Spain
Tel: +34 91 590 3638 Fax: +34 91 563 12 63

Australian Olive Association
PO Box 173, Muswellbrook, NSW 2333
Tel: + 61 (0)2 6547 9180

New Zealand Olive Association
PO Box 488, Masterton, New Zealand
Tel/fax: +64 (0)6 378 2949

INDEX

Numbers in italics refer to illustrations in the text

Abelout 34
Abou-Salt 34
acidity 46, 50
Africa 34, 78
Ajrosi 35
alternate bearing 17
antioxidants 28
apple weevil 113
Arbequina 31, 41
Argental 34
Argentina 10
Ascolana tenera 33
aspect 63-4
anthracnose 117
Australia 35-8
  early plantings in, 36
  growth of industry in, 37-8
  varieties suitable for, 80
Australian Olive Association 38

bacterial disease 115-6
baking 137
Barmaghi 35
Barnea 34, 40, *40,* 41, 71, 72, 73, 78, 102
Barouni 82
Barouni del Sahel 34
beetle, African black 113
birds 113-4
bitterness, removal of 48
black scale 111, *112*, 114
Blanquette de Guelma 34
Blanquillo 31
Blumenfeld, Gidon 40
boron deficiency 108, 117
botany 16
Bouchouk Lafayette 34
Bouteillan 34
bruschetta 134
bureaucracy 64
bushy vase pruning *102,* 103-4

Cailletier 34
Carolea 33, 76
Chemlali 34, 78
cholesterol 28
climate 56-9
cockatoo damage 113
cold pressing 45
cold traps 61
Coltibuono 73
Coratina 33
Corfolia 33
Cornicabra 31
Corregiolo 33
costs, setup 66
crushing, of fruit 44
cuisine 133ff
cultivars 21, 72-3
curculio beetle 113
cuttings 18

DA-12 76
Daphnoella 33
Daphonolia 33
David, Elizabeth 138-9
deer 115
deficiencies 117
dehydration 60
diets 26
Dikkam 35
diseases 111ff
Djalt 34
drainage 59
dressings 133-4

Empeltre 41
ETP. *See* evapotranspiration
evapotranspiration 60, 93
'extra virgin' 46, 49

Faneya 34
fatty acids 27-8
feeding 107-9
fettunta 134
financial planning & advice 65ff
flowering 57
  response to temperature & light 22
  timing of, 20
flowers 20, *21*
France 33-4, 78
Frantoio 33, 41, 71-3, *74*, 75
free radicals 29
frost 57, 61, 109
frost-hardy cultivars 57
fruit fly 112

fruit
  flesh-to-pit ratio 23
  formation of stone 23
  growth stages 22-4
  response to irrigation 59
  ripening 23-4
  yield 57
fruit-set 22
frying 136
FS-17 76

Galego 31
Gerboui 34
Gethsemane, Garden of, 9, *10*
Giaraffia 33
Gordal 31
grafting 85
grass grub 115
grasshopper 113
Greece 24, 33
'green ripe' 24, 43
Grossane 34
growing olives
  financial commitment 13
  history of, 24-5
  leaf extract 29
  timescale13-15
  varieties of, 30ff
  workload 13
growth cycle 21
  response to temperature 21

Halkidiki 33
Hardy's Mammoth 80
harvesting 43-4, 67, 120-1
heart disease 27-8
herbicide 99
history 24
Hojiblanca 31, 81
hygiene 15

income projection 67-8
infused oils 137-8
insect pests 111-2, 113
International Olive Oil Commission 25
  IOOC standards 50
  tasting terminology 51, 53
IOOC. *See* International Olive Oil Commission
irrigation 59-60, 87-8, 92-3
  and salinity 60
  and yield 109

Israel 78
Italy 31-2

J1 and J2 80
Jelin 35
Jumbo. *See* King Kalamata

Kadesh 34, 83
Kaiss 34
Kalamata *14*, 19, 20, 33, 71, 75, 81, *82*. *See also* King Kalamata
kangaroo 113
Kasb 35
King Kalamata 82
Konservolea 33
Koroneiki 33, 60, *77*, 77

lampante 50
Laudemio 73
Lavee, Shimon 40, 78
layout 87-90
leaf extract 29
leaf-roller caterpillar 114
leaves 20
Leccino 33, 41, 42, 60, 71, 75, 90
Leccio del Corno 33
Lechín de Sevilla 31
Limi 34
linoleic acid 27
Lucques 34

machinery 109-10
maintenance pruning 105
management 100ff
  in early stages 99
Manzanilla *21*, 31, 34, 40, 42, 72, 81
  as pollinator 73
marinade 135
marketing 15, 127
Maroudas, Peter 88
Marsalina 34
Mastoides 33
Maurino 33
mechanical harvesting 110, 121-2
mechanisation 25-6
Mediterranean diet 11, 28
Mediterranean fruit fly 112
Megaritiki 33
Mehravia 34
Meski 34
Meslala 34
Middle East 24, 34, 78
Mission (WA) 80

Mission 41, 42, 79
monocone *102*, 102-4
Moraiolo 33, 75-6, 96, *98*
Moresca 33
multiple trunks 19

Nab Tamri 82
Nabali Baladi 34
Nabali Mouhasan 34, 40, 78
Nevadillo 31
New Zealand 39-41
  varieties suitable for, 80
Nocellara del Belice 33
nurseries 83-5
nutrients 62

oil
  baking with, 137
  bottling 127
  classification 50-1
  cold pressed 45
  costs 68-9
  defects of, 51-2
  'flowers' 46
  frying with, 136
  infused 137-8
  matching type and purpose 54
  positive attributes 52-3
  production 118ff
  storage of, 47
  tasting and using 49ff
  varieties for, 75ff
  'washed' 46
oleic acid 27
oleuropein 24, 29, 128
olive *(See also* Growing olives*)*
  adaptation to climate 17
  as agricultural alternatives 11
  as shelter plant 60-1
  botany of, 16
  colouring of, 27
  cuisine 133ff
  eating 128ff
  establishment in Southern Hemiphere 10
  fruit fly 111
  future growing areas 12
  handling 121
  harvesting for eating 128-9
  knot 116
  moth 111
  nurseries 83-5
  oil diet 26
  oil varieties 75ff
  pickling 80ff*129*, *131*
  pressing *119*
  selecting varieties 71ff
  styles & recipes 130-2
  tree bug 112
  varieties for oil 75ff
  yields 67
Olives Australia *84*
One Tree Hill 80
organic growing 99

Paragon 80
pasta, sauce for, 135
peacock spot 78, 116
peacock's eye. *See* peacock spot
Pendolino 33, 75
Pendouiler 34
pests & diseases 111ff
pH 62
Phytophthora 116
Picholine 34, 42, 78
Picholine Marocaine 34
pickling 47-8, 130-2, 138
  timing of, 43
  traditional methods 44
  recipes 130-2
Picual 31, *76*, *77*
Picudo 31
Pigalle 34
Pinzimonio 134
planning 56, 64
planting 87ff,94-7, *95*
  alignment of rows 89
  costs 67
  layout 87-90
pollination 20, *22*
  wind and, 21
  planting layout to ensure, 90
pollinators 73-4
polyphenol 27, 119
Ponder, Mike & Diane 40, 59, 72
Pooch 80
possum 113, 115
press 45-6, *123*, *124*,
  selection of, 123
  types of, 122-4,
  setting up 70
pressing 44-7, 122ff
  rates 125
prices, for fruit 68-9
problems 111ff
processing 15

prostaglandins 27
pruning 19, 100ff *106-107*
for machine harvesting 102
maintenance pruning 105

Rakino 80
Razzaioi 33
recipes 133ff
revenue projections 68
ripeness 119
judging *23*, 48
ripping 91, *92*
Romans 25
root rot 116
root system 18
Rouget 34
Rutherglen bug 113

SA Verdale 83
salmoriglio 135
Salonenque 34
scale insect 111
seedlings 17, 18
planting 95-7
protection *96*, *98*
transporting 95
Seresin Estate (Marlborough) *12*
Seresin, Michael 41
Sevillano (Spanish Queen) 42, 71, 81
shelter 60, 61
shotberry 22
Sigoise 34
site preparation 91
site selection 55, 63
soil conditions 17, 62-3
Sourani 78
Souri 34, 40, 78
South Africa 42
South America 41
spacing of trees 88
Spain 25, 30-1, *31*, 77
Spanish Queen *See* Sevillano
spheroblast 19
spray 99
spreadsheet 65
stakes 60-1, 93-4
stock, origin of, 83
selecting 85
Stravolia 33
suckering 19
suckers *105*
Super 80

Sutter, George 36

Taggiasca 33
Tanche 34
Tanche de Nyons 81
tapenade 138
tasting procedure 53
tax 65
Tefahi 34
terroir 11, 50
Throubolea 48
timing of harvest 118-9, 128-9
Tiny Oil Kalamata 80
tocopherol 27, 28, 119
Tonda di Cagliari 33
Tonda Iblea 33
top working 85, *86*
trace elements 108
trees, selection of, 71ff
triglyceride *26*, 27

United States 41, 79, 83
Uovo de Piccione 33, 34, 40, 81

Vassiliki 33
Verdale 78
Verticillium wilt 116
'virgin/extra virgin' 46, 49
definition 50
viruses 117

Waipara Springs *39* , *40*, 41
wallaby damage 113
water 59-60
water shoot 103
weed control 91
weeds 99
wind 60-1
wood, use of, 107

Yacouti 34

Zaity 34
Zalmati 34
Zone diet 26